TRADITIONS, TYRANNY
AND UTOPIAS
Essays in the Politics of Awareness

TRADITIONS, TYRANNY AND UTOPIAS

Essays in the Politics of Awareness

ASHIS NANDY

OXFORD
UNIVERSITY PRESS

OXFORD

UNIVERSITY PRESS

Oxford University Press is a department of the University of Oxford.
It furthers the University's objective of excellence in research, scholarship,
and education by publishing worldwide. Oxford is a registered trademark of
Oxford University Press in the UK and in certain other countries

Published in India by
Oxford University Press
22 Workspace, 2nd Floor, 1/22 Asaf Ali Road, New Delhi 110002

First published 1987
Oxford India Paperbacks 1992
23rd impression 2024

ISBN-13: 978-0-19-563067-1
ISBN-10: 0-19-563067-X

Printed in India by Manipal Tecnologies Limited, Manipal

For those who dare to defy
the given models of defiance

Contents

Contents

Foreword

Occident is an accident. For the first time in human history, since what the occidentals call their 'renaissance'—that is, the simultaneous birth of capitalism and colonialism—science has been separated from wisdom and a technique has been developed for techniques.

Science has been separated from wisdom in the sense that the organization *of means* has become independent of the reflection *on ends*. In all other cultures, for example in those of India, China, Islam (so far as Asia is concerned), one recognized two uses of reason: one proceeded from cause to effect and permitted adaptation to nature, and the other proceeded from ends to ends, from intermediate ends to higher ends, and gave direction to life. Western thought has let the second use of reason atrophy. Cut off from wisdom, occidental reason has become infirm, mutilated and monstrous, indifferent to all human finality.

That which the West calls 'progress' is the hypertrophy of the first use of reason, which, as Descartes wrote, 'makes us masters and possessors of nature'. The only criterion and only value are those of 'efficacy'. Linear progress, as conceived in the West, is growing efficacy in the destruction of nature and of people.

Predictably, the Western model of growth is characterized by blind production of more and more, faster and faster, no matter what; things useful, useless or even lethal (for instance, armaments). Such 'growth' in the West is possible only by plundering the rest of the world. The 'growth' began with the genocide of American Indians, continued with the trade of African slaves, and in Asia with the opium war and the bomb on Hiroshima. This 'growth' led, in 1980, to the starvation death of fifty-five million human beings in the so-called under-developed countries, the same year that the West's politics of armaments has ended in placing four tonnes of explosives on the head of each inhabitant of the planet.

'Underdevelopment' is not a phenomenon of backwardness; it has been created by the growth of the West. The growth of some countries and the underdevelopment of others are only two faces of the same planetary maldevelopment. The first consequence, a theoretical one, of this reckoning is to denounce the falsehood which is involved in proposing to the 'underdeveloped' countries that they should imitate the Western model of development, because, by definition, a system where the growth of some countries demands the pillage and underdevelopment of three-fourths of the world is not applicable to the entire universe.

The second consequence, a practical one, is to denounce the hypocrisy of 'aid'; the interrelationship of the West with the underdeveloped countries needs to be radically altered. If the growth of the West has engendered and aggravated the underdevelopment of the rest of the world, and led to tensions and planetary suicide, the only remedy is that the West change its model of growth.

It is a murderous illusion to talk of developed and underdeveloped countries; in reality there are only the sick and the deceived. The Western countries are sick—with their blind economic growth and underdeveloped culture, wisdom and faith—which in turn breeds violence and pointless lives, incapable of proposing and realizing human ends. The deceived countries are those made to believe that their future lies in imitating the sick countries.

The principal obstacle to the necessary change is that the West, after four centuries of unshared domination during which it has exercised a disastrous impact on the planet, imposes not only its economic, political and military 'order', but also the form of culture and history which justifies it, as if the historical trajectory followed by the West was the only possible one, exemplary and universal.

The West has confiscated the universal. Starting from there it pretends to place all others on its own trajectory (a country is considered the more developed the more it resembles the West). Ashis Nandy powerfully underscores this imposture of the West when he writes: 'The modern concept of history has itself been a major means of oppression.' Therefore his effort, not to repudiate or reject the Western culture in its entirety but to relativize it, is of major importance for the fate of this planet.

In his study on Gandhi he brings out the fundamental theoretical contribution of the Mahatma: not to refuse social change, but not to confound social change with Westernization. For Gandhi, Nandy shows, to change India was not to strive to copy the West, nor to effect a 'return' to an ancient tradition. Nandy writes, 'The choice is not between a traditional technique and a modern technique; it is between different traditions of technology.' That is because no technology is neutral: it carries with it a total way of looking at and living with nature, other people and the future.

It is time to take note that the Western model of growth, which has led us to pointless existence and death, seeks to justify itself through a model of culture and ideology which carries within it the germs of death:

— an aberrant conception of nature, considered as our 'property' which we are entitled to 'use and abuse' (as Roman law defines such property), even to the point of regarding it only as a reservoir of nature's wealth and a depository of our refuse. Along that road, we are destroying our own vital environment by the thoughtless exhaustion of our resources and by pollution, and becoming unconscionable collaborators with the law of 'entropy', with the degradation of energy and the growth of disorder,

— an inhuman conception of human relationships based on unfettered individualism, that only creates societies with competitive markets, confrontation, violence and economic and political coalitions which, blind and all-powerful, enslave or devour the weakest members;

— an inhuman conception of the future which will only be an extension and quantitative outgrowth of the present, without human purpose or divine intervention, with nothing to transcend the horizon to give meaning to our lives and to turn us from the way to dusty death.

There cannot be a new world economic order without a new world cultural order. A new world cultural order will be the transition from the Western hegemony to a planetary planning for a human project. The dialogue of civilizations has become a necessity, urgent and unexceptionable. A question of survival. The central and vital debate of our times is no longer between a capitalism which generates colonialisms, wars and the ultimate crisis of our Western civilization, and a 'socialism' of the Soviet model which, by assuming the same growth objectives as the capitalist West, has become like the latter an oppressor of its own people, exploiter of the third world, and participant in the same race for hegemony and the armaments of terror. The central and vital debate of our times is the one outlined by Nandy: that between an 'alternative perspective' and 'modern oppression'

That alternative can only emerge from a free 'dialogue of civilizations' which would show how the non-Western cultures conceived and lived with other kinds of relationships with nature, with people and with the direction of their life, not for 'returning to the past', but for distilling a future out of the experience of all civilizations and not that of simply one. This is what the West has hitherto resisted, for there cannot be a true dialogue till each participant is convinced that there is something to learn from the others. And the West continues to pretend that it has a monopoly of the truth, even though it has lost the taste for researching the significance of life.

Let us be grateful to Ashis Nandy for having laid bare the errors of a 'technique for the technique' which leads the technocrat to impose on the 'profane' a system founded on the implicit postulate: 'all that is technically possible is necessary and desirable'—which leads us to actual impasses of armaments and hunger; to a technocratic medicine which treats illness as one repairs a truck, without the participation of the ill person; to the psychoanalyst who believes himself 'emancipated' in order to place himself above the patient; to the Marxist politician who firmly believes himself seized of the absolute truth, the more so for rejecting the 'spontaneity' of the masses; to the 'liberal' economist who has proclaimed for two centuries, despite all evidence and all disasters, that 'if each pursues his personal interest, the general interest will be secured', the basic principle of individualism in the jungle of capital which transforms so easily into the totalitarianism of the termites. Let us be grateful to Ashis Nandy for this vigorous contribution to the dialogue of civilizations for the unity of science and of wisdom, of politics and of its human finality.

Roger Garaudy

Preface

These essays are part of a continuous open-ended exercise. They try to spell out, with occasional help from a 'savage' reading of Freud, the basics of an alternative perspective on modern oppression. The perspective, actually the rudiments of a political psychology of man-made suffering, sees institutionalized oppression as a process ceaselessly trying to co-opt the physical and psychological worlds of the 'victorious' and 'defeated' and to destroy the basis of all dissenting visions of a more just world. This of course is the other side of the much-studied problem of legitimation, and the nature of cultural resistance or non-resistance offered by the victims of history. The essays argue that such co-optation of the outer and the inner worlds, in both the makers of history and their willing or unwilling victims, may come not only through a repressive political socialization and totalizing media but also through a worldview which denies the continuity between the oppressor and the oppressed, between the process of oppression and its understanding, and between the objective forces of oppression and their subjective counterparts. It is as a part of these denials that this worldview seeks to dismantle all readings of cultures, sciences and ethics carrying alternative visions of nature as well as of human nature, particularly all less dualistic, non-modern interpretations of modern oppression. In place of these alternative cultural theories of man-made suffering is ventured a new hegemonistic universalism, sustained by the faith that a secular, scientific understanding of history can be ameliorative or therapeutic in itself, and that the experience of man-made suffering can sometimes be a civilizing experience and a training ground for the victims. The post-Enlightenment modern world, innocent of the new forces of oppression and totalism released by modernity itself, has consistently promoted a set of secular theories of salvation which would have the oppressed share the oppressors' utopia—conservative, liberal or radical.

The following pages use some elements in the traditional visions of liberation, classical as well as 'folk', and a new reading of Gandhi to advance the view that a theory of freedom today must seriously consider and build upon the civilizational perspectives of those who, even in their defeat, even when stripped of their autonomy, dignity and means of survival, have dared to reject the values of masculine achievement, productive work and technocratic expertise to protect and nurture, however clumsily, alternative concepts of compassion, freedom, justice and dissent. The essays articulate the belief that all man-made suffering is one and the cultural and personal disintegration of the winners of the world—be they in the first world or in well-protected sanctuaries in the second or the third—is a direct result of their dominance and willing participation in a 'legitimately' inequitous and unjust order. As Romain Rolland wrote to Freud in 1923, 'Victory is always more catastrophic for the vanquishers than for the vanquished.'

Finally, the essays can also be read as an attempt to move towards a theoretical frame for understanding social intervention outside the compass of evolutionist, technocratic pragmatism. As a first step towards the development of such a frame, the essays venture an outsider's critique of the ideologies of normality, masculinity and adulthood, and a predictable critique of technological rationality, and *homo oeconomicus*, all of which have been crucial planks in the various secular theories of salvation and progress which have been popular during the last one hundred and fifty years of Western hegemony. The critiques have been offered with an awareness of some of the fundamental cultural disjunctions of the present times (for instance, between the young and the old, and the male and the female; between man and nature, and man and society; between man and his knowledge, and man and himself) and with an awareness that the time has come for us to restore some of the categories used by the victims themselves to understand the violence, injustice and indignity to which they have been subjected in our times.

There is an implicit recognition in the essays that these neglected categories provide a vital clue to the repressed intellectual self of our world, particularly to that part of it which is trying to keep alive the visions of a more democratic and less expropriatory mode of living.

To that other self of the world of knowledge, modernity is neither the end-state of all cultures nor the final word in institutional creativity. Howsoever formidable and permanent the edifice of the modern world may appear today, that other self recognizes, one day there will have to be post-modern societies and a post-modern consciousness, and those societies and that consciousness may choose to build not so much upon modernity as on the traditions of the non-modern or pre-modern world. After all, modernity itself drew more upon the classical Hellenic than upon the medieval Christian traditions.

Thus, the flip-side of any cultural self-exploration outside the West today has to be an archaeology of knowledge which excavates and fights for a lost or repressed West. Knowledge, too, like suffering, is an indivisible human experience. Self-aware, self-critical knowledge has to realize its own indivisibility by reaffirming the indivisibility of human and social choices in the matter of human happiness and suffering and human ends and means.

A preface can also be an apologia. I shall take advantage of that right to point out what the discerning reader will find out in any case—that I am neither a political philosopher nor a philosopher of science. I have reached the position marked out in the following pages in the course of a self-exploration while working in two overtly unrelated areas: political psychology and cultures of science. The first area made me aware that there is not only a psychology of politics but also a politics of psychology. The second made me aware that there is not only a culture of science but also a culture in science. The first led me from the science of politics to the politics of science; the second from the culture of politics to the global politics of cultures, within which the culture of modern science seems to provide the basic format of domination. These essays, therefore, have helped me reconcile two long-term concerns of mine and to reconceptualize the political relationship between actors and texts. This book is a direct result of the awareness that there is an isomorphism in the internal structures of knowledge, persons and cultures (that is, crudely speaking, what we do to others we do not only to ourselves but also to our cognitive ventures) and that power is the 'legitimate' modern means of denying this isomorphism to ourselves and forcing this denial down the throats of others (because reportedly progress and realism demand this denial).

This awareness has forced me to sympathetically re-examine some aspects of the nonmodern visions of a good society. Seemingly these visions convert a part of the drive for power to a drive for power over self, particularly over the unacceptable parts of one's self. Many of us have learnt to see such 'private' power either as useless or as a defensive stance compensating for real power in the real world. Yet, while justified in terms of the needs of the spirit, such a drive for inner power has social-critical functions unacknowledged in most modern theories of legitimation. It denies absolute value to any secular theory of society and it allows some knowledge to be subversive by allowing the users or producers of knowledge to take advantage of the contradiction between the outer and the inner powers. I have not ignored the so-called real power in the real world in these pages, though I have tried to show that it is often not so real after all. But I hope I have restored something of the dialectic between outer and inner powers by showing that the so-called unreal inner power, too, often enough, is not so unreal after all, even by the strictest secular criteria. This book, thus, takes to a conclusion the argument of my earlier book, *The Intimate Enemy*.

Essays such as these cannot but change over their various versions, in response to criticisms, suggestions and encouragement. Thus, the forums which allowed me to present my ideas and to debate or discuss them with scholars from all over the world have contributed immensely to the present forms of the essays.

'Evaluating Utopias' and 'Towards a Third World Utopia' were written during 1978–81 for the group on Alternative Visions of Desirable Societies. The group was set up by the Centre de Estudios Economicos y Sociales del Tercier Mundo, the World Future Studies Federation, and the United Nations University. A part of 'Evaluating Utopias' was published in *Mazingira*, 1980 (12) and a later version in *Seminar*, January 1984 (250). Earlier incarnations of 'Towards a Third World Utopia' were published in *Alternatives*, 1978–79, 4(3); and in Spanish in Eleonora Masini and Johan Galtung (eds.), *Visiones de Sociadades Deseables* (Mexico City: WFSF and CEESTEM, 1981). A revised version was published in Eleonora Masini (ed.), *Visions of Desirable Societies* (London: Pergamon, 1983); *Problems of Non-Alignment*, 1984, 2(3); and in Jan Danecki (ed.), *On Mal-Development* (forthcoming). Abridgements or extracts were published in R. B. J. Walker (ed.), *Culture, Ideology and World Order: Essays in International Political Theory* (New York: Westview Press, 1984); Thomas Pantham and Kenneth Deutsch (eds.), *Political Thought in Modern India* (New Delhi: Sage, 1986); and in *The Ecologist*, 1983, 13(4).

'Reconstructing Childhood' has grown mainly out of a lecture delivered at the Conference on Culture as a Social System, organized by the National Institute for Research Advancement and Tsukuba University at Tsukuba, Japan, in November 1981. Earlier and briefer versions of the essay were published in *The Times of India*, 2–4 February 1982, and in *Resurgence*, 1982 (92). The present version was written for *Alternatives*, 1984, 10(3).

'The Traditions of Technology' originated as a brief note written for a seminar on Traditional Technology and the Poorest Sections of Rural Communities, held at Kathmandu in October 1977 and organized by the Quaker International Affairs Programme. Later versions were published in *Alternatives*, 1978–79, 4(4); Ward Morehouse (ed.), *Science, Technology and the Social Order* (New Brunswick, N.J.: Transaction Books, 1979); *Man–Environment Systems*, 1984, 13(6); and in the form of an interview, 'Dialogue on the Traditions of Technology', in *Development*, 1981, 3(4).

'Science, Authoritarianism and Culture' is a reincarnation of the M. N. Roy Memorial Lecture, 1980, delivered at the Gandhi Peace Foundation, New Delhi, on 21 March 1980. The lecture in its original form was published in *Seminar*, 1981 (261), and extracts from it were published in *Radical Humanist*, May 1980; *Resurgence*, 1980 (83); and *The Ecologist*, 1983, 13(6).

'From Outside the Imperium' was presented at the meeting on Culture, Power and Transformation, organized by the World Order Models Project, and held at Lisbon in May 1980. A crude, early version was also published in *Alternatives*, 1981, 7(2). The framework of this version was presented at a meeting organized by the Lokayan group at Delhi in 1982. Parts of it were published in *The Economic Times*, 2 October 1983; and in Ramashray Roy (ed.), *Contemporary Crisis and Gandhi* (forthcoming).

For criticisms, suggestions and editorial help, I am grateful to the participants in the meetings at which the papers were presented, to the editors of the journals in which they appeared, and to the numerous friends and colleagues who have commented on the papers. All of them share a part of the responsibility for the content of these essays, including the faults which persist in them. I must particularly mention Giri Deshingkar, R. A. P. Shastri, Girdhar Rathi, Claude Alvares, M. P. Sinha, Eleonora Masini, Shiv Visvanathan, Johan Galtung, C. R. M. Rao and Mira Sinha.

I must also thank Roger Garaudy for his Foreword, and Manish Nandy for translating it from the French.

Evaluating Utopias: Considerations for a Dialogue of Cultures and Faiths

I

Whenever men have taken utopian descriptions seriously, the result has been disastrous.

Jacques Ellul[1]

When the gods want to punish us, Oscar Wilde might well have said, they grant us our utopias. A realized utopia can be another name for terror. Most utopias are a threat not only to their detractors, but also to their own partisans. One doubts if some of the great utopia-builders would have been as willing to live with their models as their disciples frequently are. Even though Maurice Merleau-Ponty says so in an altogether different context, his cynical comment has been internalized by many utopians: 'Once humanism attempts to fulfil itself with any consistency, it becomes transformed into its opposite, namely, into violence.'[2]

A minor tragedy? Or a safeguard against self-destruction, as Wilde perhaps would have argued? Can it be that utopians have often unconsciously feared their own utopias and that the hypocrisy, dishonesty and betrayal of cause we often find in them is only an unwitting means of subverting their own creations?

One feature of the popular visions of the future hints at an answer: rarely have utopians and visionaries built escape clauses

[1] Jacques Ellul, 'Search for an Image', in Robert Bundy (ed.), *Images of the Future: The Twenty-First Century and Beyond* (New York: Prometheus Books, 1976), pp. 24–34; see p. 25.

[2] Maurice Merleau-Ponty, 'Koestler's Dilemmas', in Murray A. Sperber (ed.), *Arthur Koestler: A Collection of Critical Essays* (Englewood Cliffs, N.J.: Prentice-Hall Inc., 1977), pp. 69–85; see p. 78.

into their charters for the future. One can enter their utopias; one cannot emigrate from them. Most utopias fear thoughts which may sabotage them, features which acknowledge their mortality, and situations which may make them irrelevant. Alternative utopias and worldviews they see either as conspiracies or as the ideologies of conspirators. Utopias are loath to grant that they may be made irrelevant by new readings of old visions or by new visions with changed perceptions of evil. Some sophisticated utopias do anticipate attacks from outside, the subtheories of counter-revolution in Leninist Marxism and in contemporary revolutionary Islam being good examples. But such anticipations are framed in the totalism of the existing utopias, so that even in defeat no important aspect of these utopias is negated; so that, even when times change and followers defect, one can hold on to one's faith and accuse the defectors of being dishonest or conspiratorial. This not merely protects one's self-esteem but also guards one's worldview against the ravages of time. Our immortality lies in our visions; admitting their mortality or irrelevance is admitting our own mortality and irrelevance.

Robert Nozick has argued in his work on collective choices that a universally-shared utopia is not really possible.[3] I am not sure if he would grant the possibility of partly-shared cultural utopias; his choice-makers are basically cultureless, ultra-rational robots, trying to live up to their creator's concept of human beings standing outside time and space. They seem to have no awareness that our times have seen both the destructive and the constructive pulls of shared cultural visions, irrespective of their rational or logical status. Neither the unworkability of utopias nor their internal inconsistency has ever detracted from their power. Indeed, such unworkability and inconsistency have often given a utopia greater appeal and longevity by guarding it against uncomfortable realities and by allowing for diverse, conflicting interpretations of its text.

Perhaps utopias humanize or dehumanize us by intervening

[3] Robert Nozick, *Anarchy, State and Utopia* (Oxford: Basil Blackwell, 1974) Part III, especially chapter 10.

in the contemporary social consciousness, not by producing the blueprints of a realizable future world.[4] Their strengths as well as weaknesses lie here, not in the societies or human beings they promise to produce. Utopias are always real for those who live by or with utopias, either as participants or as victims. It is thus that utopias reconcile ends and means, and subjects and objects. Much before our utopias can independently oppress anyone, we oppress through and are oppressed by our utopias. Much before we can control the future through our utopias, the future becomes our tool to control others *now*. The quality or practicability of utopias is not a relevant concept here. Indeed, an overly determined attempt to actualize a utopia can turn it into a dystopia for many or destroy its 'pull' by exposing it to the harsh light of human experience.

Perhaps a part of the power of our visions comes from their very unrealizability—from their impractical, 'utopian' scaffolding and from their implicit, unattainable, normative codes. It is a creative tension with which some persons and cultures prefer to live. The gap between reality and hope which such a vision creates becomes a source of cultural criticism and a standing condemnation of the oppression of everyday life to which we otherwise tend to get reconciled.

Organized ideas about utopias, therefore, are like sacred texts in plural cultures; you read into them—or read out of them—according to the needs of a person, institution, culture or an age. Because the needs lie in the present—and because a utopia survives and dies within the minds of men and not in the pages of a scientific treatise—the creative and destructive powers of a utopia, too, are mostly located in the present. A utopia is to be understood in the context of the strengths and the crises of the way of life in which it is rooted.

Hence, the first question for those interested in a dialogue of utopias and faiths is not how to combine the best of different utopias we live with, so that a single, concrete, rational, realizable utopia emerges in the future. No utopia can integrate themes or modules borrowed from outside its boundaries on

[4] See 'Towards a Third World Utopia' in this volume.

purely intellectual grounds; a vision is not a scholarly study, though it can be the subject of one. It is part of a life-style; it is informed by a certain faith and passion.

The first question, therefore, is how each utopia can legate with others to enrich itself and the others, to deepen the lives of its partisans and make them aware of their own long-term social goals and strategies (through a process akin to therapeutic intervention in depth psychology). This calls for an assessment, rather than an integration of visions. Modern communications and growing cross-cultural contacts have already managed to ensure, in crucial areas of life, the formidable presence of a single utopia: the One World which nineteenth-century Europe visualized. To endorse that presence is to further marginalize cultural dissent and to further abridge cultural plurality. At a time when most visions are struggling for survival, a dialogue of visions must first be a statement against uniformity. This may even involve the responsibility of leaving out of the dialogue some of the more fragile or experimental utopias, that of a small tribe here or a commune there, which are immensely valuable in themselves but are incapable of withstanding aggressive external evaluation. I am here primarily concerned with the sturdier breed of visions of desirable worlds which are always there, breathing down the necks of our existing utopias as possibilities, alternatives or threats, or as 'trips' which we would be tempted to take if only we knew the safety margins.

Which utopias, then, to bring into a dialogue? Which utopians will be interested in a dialogue? Would either Christianity or Marxism, for all their ongoing exchange, be willing seriously to relate to the articulate, once-influential utopia of National Socialism, however deeply rooted the latter might be shown to be in some aspects of European culture, or in the unromantic, murderously practicable versions of the romantic social visions which became popular in Europe in the late thirties? Or would Christianity and Marxism only wish to learn from the experience of—and with—the National Socialist vision?

Perhaps some utopias can never be brought into a dialogue; they can only be the themes in a dialogue. Perhaps there is a sociometry of utopias, a matrix of utopias-as-neighbours and utopias-as-strangers. This sociometry defines for each utopia its reference group of other utopias which pose challenges and invite self-exploration, and a reference group of dystopias or anti-utopias which can be objects of study or debate and sources of anxiety or fear but not of dialogue. Seemingly, only the utopias shape human fate and the course of history. Yet, the dystopias are the ones which in our times have shown especial resilience in the minds of men, both as negative experiences and as fictional warnings. They may not always be a part of our positive human heritage but they always are a part of our human selfhood. Our hells may be paved with our utopias but our heavens are sign-posted with dead and living dystopias.

This subjective sociometry of utopias also includes a few competing utopias or counter-utopias as hidden reference points. Self-consciously or not, we are often in dialogue with utopias which define, or once defined, our own utopias as a starting point or as a target of defiance. For instance, classical or greater-Sanskritic Hinduism, heavily dependent on Shankaracharya's metaphysics of *advaita* since the eighth century, continues to see Buddhism as a competing counter-utopia as well as a major source of ideas and values, more than one thousand years after Buddhism had apparently lost the battle of minds in India. This perception is not conscious in even the self-conscious Hindu, but it is woven into the structure of his religious and social life. This is precisely because *advaita* was a response to Buddhism and helped Hinduism to internalize major Buddhist categories, so that Buddhism no longer remained only an external challenge but became an internal criticism that had to be grasped by each Hindu vision of an ideal society. Elsewhere I have argued that a similar process was unleashed in India by the modern West during the British colonial period. So that while even the 'universal' Western visions of a good society do not have to include India, except peripherally as an object of study, charity or experimentation, even the more parochial Indian visions have to include some version of the West, either as an internal ally or as a critic.[5]

II

What if we want to transcend a utopia's inner criteria to assess other visions of desirable societies? Can we construct a set of criteria to critically assess others' utopias without being weighed down by the prejudices of our own? The answer will have to be based on a new awareness of our experience with faiths in this century, especially with the new violence which modern institutions and ideas have released. This century has taught us that the search for a non-oppressive society can itself sometimes become a new means of oppression and a technique of expropriating new kinds of surplus—economic, political, cultural and psychological. Our criteria for evaluating utopias must include safeguards against the criteria themselves.

The first and most important of these criteria has to be the ability of a utopia—or a vision or a faith—to be accountable for its legitimate and illegitimate brain-children. Partisans of a utopia tend to shed responsibility for crimes committed in its name, while taking full credit for the good.[6] They try to cope with this split (1) by rereading the utopia and coming out with a 'purer' or 'truer' version allegedly unavailable to its earlier misguided partisans and (2) by redefining the partisans who 'misuse' the utopia as false or motivated partisans who in reality are no partisans at all. Thus, millions have been killed as enemies of Christianity, millions have been enslaved so that they get a Christian civilization, millions have been accused of paganism and then denied the right to survival. And all this has often been explained away as deviations from true Christian ethics and as the work of Christians who did not know their

[5] Ashis Nandy, *The Intimate Enemy: Loss and Recovery of Self under Colonialism* (New Delhi: Oxford University Press, 1983), chapter 2.
[6] Elsewhere I have called this the process of split or schizoid legitimation. It involves contextualizing the social evils associated with a vision, and decontextualizing the good.

faith, so as to allow the Christian vision to emerge from history innocent and uncontaminated.

Similarly with the blood-stained, oppressive heritage of a number of oriental religious ideologies, now seemingly so innocuous in their powerlessness and immaculate in the hands of their contemporary interpreters. It is possible to construe a Hinduism which sees untouchability as an aberration or as an accident of history; it is possible to construe a Hinduism which sees untouchability as a necessary product of some parts of Hindu cosmology. The choice is metaphysical *and* political.

Likewise, there can be a Marxism which considers Stalin an anti-Marxist sadist, and a Marxism which acknowledges, as part of its self-evaluation, those aspects of Marx's conceptual apparatus and methodology which legitimized Stalinism as a necessary part of Marxism. Oppression can be, and sometimes is, unintended, but no theory of oppression is complete unless it admits that it is determined, space- and time-bound, and flawed. This admission is the obverse of Hannah Arendt's attempt to locate the sources of institutionalized violence in world-views and ideologies rather than in intentions or motives.

Perhaps a vision should be judged not so much by what is done in its name as by its ability to sanctify accountability and self-exploration. Not the self-exploration of extracted confessions which adorns the history of the late medieval witch-hunt in Europe, nor its modern version tried out in Soviet Russia in the thirties. But the self-awareness from which the visions of a plural, more humane polity might emerge to transcend the chilling record of violence inflicted in the name of utopias and the even more chilling record of state-sponsored self-condemnations associated with so many attempts to translate utopias into practice. No utopia can give a guarantee against its misuse by over-zealous ideologues, but an utopia can build conceptual components which sanctify self-doubts, openness and dissent.[7]

[7] Johan Galtung's demand that every vision of a desirable society must include within it built-in contradictions is obviously designed to subserve the same purpose. 'Alpha and Beta and Their Many Combinations', presented in the first

Second, a utopia must be able to take criticisms from other utopias as if the criticisms were partly undetermined by social, political and psychological forces. And it must be able to view its own criticisms of other utopias and visions as at least partly determined by interests and/or drives. I like to believe that this principle is Gandhian. In the 1920s, when feelings were running high in British India because of the freedom movement, Catherine Mayo wrote her savagely anti-Indian and pro-imperialist treatise, *Mother India*.[8] Gandhi called the book a 'drain inspector's report' but added that every Indian should read it. While Mayo's critique of Indian culture was blatantly prejudiced, he seemed to imply, Indian culture should have the self-confidence to put her criticism to internal use. After all, the Mayos are transient phenomena; cultural renewal through internal criticism is a more serious, long-term affair.

In this respect, Ananda Coomaraswamy is the flip-side of Gandhi; he refused to use any aspect of even a nativized version of modernity as an indirect source of social criticism. His attitude to caste is an example. Crudely speaking, caste to him was an arrangement not unique to Hinduism; it is, or was, found among the Iranians, the Japanese and the Hebrews and was common in feudal Europe. The hostility of the modern world to caste, Coomaraswamy argued, was basically the hostility of an industrial society to a pre-industrial one, because the status and dignity granted to the low castes in traditional societies was greater than that granted to the industrial proletariat in the modern West.

Empirically, Coomaraswamy could be correct. Also, as a pluricultural Sri Lankan Tamil, fighting the self-hatred and intellectual dependency of the colonized world, he had the right to be defensive. Today, an Indian interpreter of the traditional Indian vision of a desirable society, like any serious political or social activist, has to be answerable to Indians surviving in a political and social reality called India, not merely to a rarefied

---

meeting of the group on Alternative Visions of Desirable Societies, Mexico City, 1978.

[8] Catherine Mayo, *Mother India* (New York: Harcourt, Brace, 1927).

idea of Indianness. He must own up, on behalf of his culture's vision of the future, the oppressive contents of his culture in the present. The battle fought over the last one hundred and fifty years in India by social and religious reformers, many of them intellectually far humbler than Coomaraswamy but as deeply rooted in the living culture of India, is an indirect admission of this responsibility. This responsibility cannot be avoided in the name of cultural relativism, even in those post-colonial societies which have seen the oppressive contents of traditions being used as a justification of colonial rule. In any society one has to own up, as one disowns.[9]

Only by retaining a feel for the immediacy of man-made suffering can a utopia sustain a permanently critical attitude towards itself and other utopias, and yet have a creative dialogue with the latter. A utopia is a language; it is a language of interpretation and criticism, an 'exercise in suspicion', as Paul Ricoeur calls such interpretive exercises.[10] Neither the criticism nor the suspicion can ever end. There are always interpretations of interpretations, and meanings of meanings. What some traditions call *maya* and what the moderns like to call ideology or ego defence can never be fully eliminated; its changing forms can only be seen through or demystified at different points of time.[11]

Third, a utopia must show some capacity to liberate the utopians from its own straitjacket. This paradoxical proposition links the choice of visions to the problem of individual freedom. A vision which activizes those aspects of human personality which constantly seek certainty—that is, a vision which links up with the unconscious defences of mind which endorse every search for certainty—is likely to push the human mind towards totalism. A utopia should be able to free its partisans from its own patterns and to survive in them only as a certain quality

[9] Given that Coomaraswamy was defending defeated civilizations, his reverse discrimination is of course more forgivable than Mayo's paean to European ethnocentrism.

[10] Paul Ricoeur, *Freud and Philosophy: An Essay on Interpretation* (New Haven: Yale University Press, 1970).

[11] Alan Watts, *Psychotherapy East and West* (New York: Ballantine, 1969), p. 21.

of thinking and living, as a form of suspicion of existing patterns.[12]

This is not a plea for idea-systems which have the principle of falsifiability built into them. That principle is an internal standard which specifies what a given system of knowledge at a given point of time should contain. It does not allow for the falsification of a faith on contextual grounds, for instance on the ground that the faith justifies exploitation, cruelty or authoritarianism. Conventional falsifiability is also acultural; it allows for the rejection of individual theories, not of the ethos of knowledge from which the theories come.[13] I have in mind visions which include an element of self-destructiveness. These are visions which tend to self-destroy when used against demands for justice, compassion and freedom. Some versions of the sufi concept of a good life are as good an example as any.

Of all the utopias which threaten to totalize the human consciousness, the most seductive in our times has been the one produced by modern science and technology.[14] The 'normal' culture of modern science visualizes the human future in terms of the expansion of human skills, perceptions and power through technology and, as human beings often tend to lag behind the machine in these areas, in terms of the gradual substitution of such laggards by machines.[15] Such a vision views humanness

[12] The reader may recognize that some meanings of the Buddhist and Hindu concepts of *nirvana*, *mukti* and *moksha* are compatible with such a position. For a very general introduction, see Watts, *Psychotherapy East and West*, chapter 3.

[13] As a popular novel puts it, 'The true system, the real system, is our present construction of systematic thought itself, rationality itself, and if a factory is torn down but the rationality which produced it is left standing, then that rationality will simply produce another factory.' Robert M. Pirsig, *Zen and the Art of Motorcycle Maintenance: An Enquiry into Values* (London: Corgi, 1976), p. 94.

[14] The reference here is to the visions of an ideal society associated with modern science and technology, not to modern science and technology as such. For a more detailed critique from a futuristic and utopian point of view, see Ashis Nandy, 'Science in Utopia: Equity, Plurality and Openness', *India International Centre Quarterly*, 1983, 10(1), pp. 47–59.

[15] See 'The Traditions of Technology' in this volume. This theme has been discussed in detail by Theodor Roszak, *Where the Wasteland Ends* (Garden City, N.Y.: Doubleday, 1973); and Joseph Weizenbaum, *Computer Power and Human Reason* (San Francisco: W. H. Freeman, 1976).

itself as superfluous and contaminating, and it has to try to free the future from the vagaries of the untutored imagination of the unscientific laity. It shields itself from all criticism by turning utopia-building itself into a science,[16] and promotes a search for certitude and technical finish to ensure that our visions are shaped by the existing level of technology in future studies, not by our imperfect, labile, political and social consciousness. This makes it doubly painful for the faithful to break out of their visions. Losing faith in a 'scientific' utopia appears dangerously like giving up one's security, knowledge and sanity. It seems to redefine disciplined freedom as habit-forming captivity.

Fourth, no dialogue is possible with a utopia claiming a monopoly on compassion and social realism, or presuming itself to be holding the final key to social ethics and experience. Such a vision not merely devalues all heretics and outsiders as morally and cognitively inferior, it defines them as throw-backs to an earlier stage of culture and history, fit to be judged exclusively by the norms of the vision. Thus, if paganism is an early stage of monotheism, and ahistoricity that of historicity, the monotheists and the historically-minded can not only claim to understand themselves and their world better than the primitives understand theirs, they can also claim to understand the primitives and their world better than the primitives themselves do. Indeed, as historians to the world, the historically-minded can claim to know the future of the pagans and the ahistoricals better than the latter, for that future can be no different from the present of the civilized. Both the present and the future of the savage are thus hegemonized. And the savage becomes, by definition, a motivated or misguided interpreter of the future, in league with larger forces of darkness.

Implicitly some visions see other visions not merely as competing ideologies but as conspiracies against human reason and

[16] For a critique of such utopias from a different vantage ground, see Mihailo Markovic, 'Scientific Predictions and Visions of the Future', presented in the second meeting of the group on Alternative Visions of Desirable Societies, Mexico City, 1979.

values. A dialogue with such hegemonic, parochial visions may become an invitation to ethnic suicide. The proselytizing visions especially, even when they are secular, have a tendency to devour other utopias, paradoxically by rejecting the otherness of the latter and by 'accepting' them as earlier stages of the evolution of the self. Earlier such destruction-through-assimilation was brought about in the name of revelation; these days, it is brought about in the name of revolution, development or science.

Perhaps fully organized, completely self-consistent, aggressively proselytizing, entirely self-sufficient visions cannot be brought into a dialogue. Being languages, visions have implicit definitions of oppression and liberation, privilege and under-privilege, rulership and subjecthood written into them.[17] These definitions interlace with another set of concepts of liberators, prophets, vanguards, and true and false analysts of history, texts and 'reality'. The moment one gives such visions finality and views them as sciences of human goals and as techniques of reaching the goals, the concepts subsumed under the visions acquire fixity of meaning and priority over the human beings and human conditions they are supposed to serve or describe. All dialogue among such sciences of the future is reduced to a zero-sum game, with each side seeking final victory over others, for the sake of an increasingly reified idea of the human future.

Fifth, utopias in dialogue have to be partly independent of history. All utopias are responses to human experience and, particularly, experienced predicaments. Only a few have the courage to admit that such experiences can become rigid responses to shared traumas, and can begin to restrict choices. Even fewer would admit that the past can come back only in the mind of the present, and the future, in so far as it lies in our choices today, can open up new possibilities informed with but unburdened by the past. A utopia responding directly to history either dies a natural death after it has sensitized the public realm to the historical problems which first threw it up, or it continues to limit social choices by defining the future in

17 See 'Towards a Third World Utopia' in this volume.

terms of the past. Yesterday's dissent is often today's establishment and, unless resisted, becomes tomorrow's terror.

This is not a mechanical criticism of historicism, or of those twenty-first-century visions responding to nineteenth-century predicaments. This is to recognize that utopias pre-defined by history discriminate against those who are the new victims of history. For history as a discourse is a modern medium. Utopias participating in that discourse have to be definitionally located in the worldview of the dominant; even while trying to speak for the dominated, they have to define freedom in terms borrowed from the dominant. Such utopias could hand over the world to a vision of the future which is but a corrected version or a linear projection of the present state of the privileged world.

Sixth, we are better off with negatively defined utopias than with positively defined ones. A utopia which rejects technicism or the idea of mastery over nature is often a more serious affair than a utopia which recommends a specific technoeconomic or ecological solution. Whatever be the present status of the philosophical debate on negative versus positive freedom, in recent years dystopias have survived in human minds better than utopias. The latter raise problems of conflicting values; the former promotes a vague, implicit negative consensus on an unheroic vision of a 'decent society' which may not fulfil everybody's desire for a positively designed utopia but may help the diverse concepts of a tolerable society come closer to articulation.[18] It is easier to establish communications among social criticisms than to summate the values of diverse civilizations.

III

What about the process of dialogue? Are some dialogues themselves hierarchical and potentially ethnocidal?

[18] The concept of 'decent society' is broadly that of Barrington Moore, Jr., 'The Society Nobody Wants: A Look Beyond Marxism and Liberalism', in Kurt H. Wolff and Barrington Moore (eds.), *The Critical Spirit: Essays in Honour of Herbert Marcuse* (Boston: Beacon Press, 1968), pp. 401–18.

According to the dominant format, a dialogue between two visions is established when one of them is seen as the framework, tool or theory for understanding the other, serving as the object of interpretation and as a reservoir of implicit or latent insights which could be useful or enriching for the former. For instance, during the last fifty years, highly suggestive work has been done by some physicists and philosophers to demonstrate the compatibility between ancient religious thought and modern physics. In social theory, too, such a dialogue has been established between some readings of ancient faiths and contemporary theories of social intervention. Certain schools of Marxism, for instance, have rejected the Marxist orthodoxy which sees an essential opposition between Marxism and Christianity, and accepted parts of Marxism as relevant and some aspects of Christianity as acceptable. By emphasizing a re-prioritization within the two faiths, the schools hope to use the ethical and radical potentials of Christianity in terms of their reading of Marxism.

One must, however, be aware that this approach often draws upon the arrogance of the modern world and ethnocentrism towards non-modern cultures and past times. The approach assumes that if one examines an ancient, marginalized vision in terms of a modern one and shows the former to be radical, scientific or rational, one has established a dialogue between the two. For this approval, the dialogue turns regressive if the interpretation proceeds the other way—from the vantage ground of an ancient faith (say Islam) to a contemporary one (say Marxism). Even though Islam has been with us for one and a half millennia and Marxism for one hundred years, present-day theories of progress do not favour dialogical processes which would subject the various interpretations of Marxism to a critical examination from the point of view of a new interpretation of Islam. There is a pecking order of cultures in our times which informs every dialogue of cultures, visions and faiths and which tries to force the dialogue to serve the needs of the modern West and its extensions within the non-West.

Under every modern dialogue of visions lies a hidden dialogue of unequals, with its own bevy of prosecuting and defending attorneys, witnesses and false witnesses, and partial and impartial judges. Overtly, a dialogue is supposed to involve parallel but interrelated processes of self-confrontation in each culture or faith participating in it. But in this age of the 'neurosis' of individuals and 'false consciousness' of groups, there is a well-structured politics of self-confrontation too. In that politics, a dialogue can become a technology to induce self-confrontation in the others and to make that self-confrontation the centre-piece of the dialogue. To facilitate that process, there are the modern father-confessors or psychotherapists—standardized, substitutable technocrats who have reportedly gone through the major part of their own self-confrontation and have acquired the right to professionally guide the self-confrontation of others.

Such a dialogue endorses the continuity between subjecthood in research and subjecthood in the politics of cultures.[19] Like patients under therapy and the laity under confession, a culture under study, too, has not only a worldview but is also subjected to a worldview. A culture with a developed, assertive language of dialogue often dominates the process of dialogue and uses the dialogue to cannibalize the culture with a low-key, muted, softer language of dialogue. The encounter then predictably yields a discourse which reduces the second culture to a special case—an earlier stage or simplified version—of the culture with the assertive language of dialogue.[20]

[19] This theme has been developed along somewhat different lines in Ashis Nandy, 'Towards an Alternative Politics of Psychology', *International Social Science Journal*, 1983, 35(2), pp. 323–38. For an example of such an effort, elegant but inhuman and ethnocidal, see V. S. Naipaul, *India: The Wounded Civilization* (London: André Deutsch, 1977). That Naipaul apparently uses empirical methods and once at least falls back upon psychoanalytic terminology only proves, if such proof was necessary, that modern social sciences, too, can be mobilized as witnesses for the mock trial of cultures.

[20] Once again, the process of dialogue here resembles some forms of psychoanalytic process in which the discourse which emerges from the encounter between the analyst and the analysand reduces the latter to a special case in a classificatory system of case histories. See Nandy, 'Towards an Alternative Politics of Psychology'.

One such encounter of cultures was the nineteenth-century discovery of India by European Indologists. Well-intentioned and often overtly wedded to the concept of cultural relativism, many of these Indologists contributed to the growth of a new dialogue between the East and the West. Yet the dialogue merely helped integrate India into the Western worldview, so that both for European imperialists and anti-imperialists—and for most of their Indian collaborators and imitators—every feature of Indian culture became an object of evaluation in terms of the dominant political, social, intellectual and aesthetic norms of West Europe. Thus the European rediscovery of India made it more difficult, even for Indians, to critically evaluate Indian civilization in terms of an enriched, re-examined version of Indianness. The choice became one between Westernization and static, close-ended, reactive Indianness, and the latter was set up to lose. A battle of minds is often won not through the suppression of ideas, but through a dialogue.

Perhaps the defeat of Indian cosmology in the nineteenth century was not only due to the fact that the other participant in the dialogue was, by proxy, the modern West. In practice most cultures can be totalizing; in the worldview of a complex, self-confident culture, other worldviews often become special cases whose oddities could be explained in terms of native categories. The imperial West only managed to back up its worldview in the Asian and African contexts with the help of its secular power and its modern technology.

When two cultures of unequal secular power enter into a dialogue, a new hierarchy inevitably emerges, unless the dialogue creates a shared space for each participant's distinctive, unstated theory of the other cultures or, in its absence, each participant's general theory of culture. The concept of cultural relativism, expressed in the popular anthropological view that each culture must be studied in terms of its own categories, is limited because it stops short of insisting that every culture must recognize the way it is construed by other cultures. It is easy to leave other cultures to their own devices in the name of cultural relativism, particularly if *the visions of the future of* these other cultures have already been cannibalized by the worldview of one's own. It is less easy to live with an alien culture's estimate of oneself, to integrate it within one's self-hood and to live with that self-induced inner tension. It is even more difficult to live with the inner dialogue within one's own culture which is triggered off by the dialogue with other cultures because, then, the carefully built cultural defences against disturbing dialogues—and against the threatening insights emerging from the dialogues—begin to crumble.

Thus, we come full circle. A dialogue of cultures—or of utopias, visions and faiths—is a dialogue within each participating culture among its different levels or parts. This second dialogue could be articulate, well-defined, or central to the culture; it could be inarticulate, ill-defined or marginal. Some cultures hide their most profound experiences at their peripheries but all cultures have the capacity to use creatively the intersecting demands of such outer and inner dialogues.

IV

Nobody has ever lived in a utopia. But most of us live by our utopias, explicit or implicit. We are sometimes warned that many Asian and African societies do not have clear-cut ideas of the future. As far as Asia is concerned, it is said, the emphasis is too much on self-realization and on the quest for an ideal self. It is, however, not convincing that philosophical and/or religious goals of perfection, however fervently focused on the perfection of the self, could exist in a vacuum, that is, be unrelated to a quest for a more perfect society, or to the expression of alienation from an existing society.[21] It is easier to believe that each eupsychia (the idea of an ideal personality and the idea of an ideal shared consciousness) is embedded in a utopia and each utopia in turn is embedded in an eupsychia. A utopia has to have its implicit concepts of human personality and

[21] Samson Knok, 'Form or Content? Reflections on the Concept of Utopia in Asian and West European Thought', *Alternative Futures*, 1980, 3(3), pp. 3–14; see pp. 3–4.

person-in-society. If in some societies the utopias seem ill-defined or underdeveloped, in others the eupsychias seem impoverished. A creative dialogue of visions of good societies may well begin with a dialogue between social and psychological utopias, between the visions which stress the social–environmental determination of a good society, and those which stress human consciousness as the arena where the battle against man-made suffering must first be fought.

As far as Africa is concerned, one simply does not know. Academic anthropology has not given one a straight answer and it is not clear that the answer, had it been given, would have been a help. However, I am unable to believe that a culture can live with a present that does not include, even in concepts such as that of a timeless time, at least some intimations of a vision of an ideal society mirroring a cluster of hopes and values. Such hopes and values can be identified and used as a critique of contemporaneity. A utopia can be public and grand; it can be implicit, private and modest. Even in the latter form, it can communicate with other utopias nurturing other hopes and values, and have a dialogue with a part of itself.

A dialogue with such low-key visions requires the capacity to listen with what some psychoanalysts call the third ear. Without that apocryphal ear, visions of the peripheries of the world will always seem trivial or ephemeral. The visions of the weak, when not cast in the blood-curdling language of the professional revolutionary, tend to be undervalued. Their apparent fragility, their apparent inability to withstand analytic thought, and their defensiveness and diffidence in the face of Cartesian categories—all contribute to the undervaluation.

Another proposition concerning a dialogue of visions can probably be written now: all dialogues between utopias or faiths are richer to the extent they link the simple visions of decent society and the 'grander' visions of positive freedom or of total reconstruction. Such linkage ensures not so much a dialogue between the static or reformist visions and the revolutionary ones as a dialectic between the folk and the classical, the experiential and the meta-theoretical.

Finally, the matter of dialogue with utopias which are located, paradoxically, in the past. Speaking of the simple Chinese utopias Forest Lin makes the curious point that utopias in the Chinese tradition are hamstrung by their faith in a golden age when society was allegedly perfect. Because the golden age was simple, Lin says, Chinese utopias are also simple.[22] Similar arguments reappear in other forms whenever the modern student of utopias confronts a culture speaking in the metaphors of a past that refuses to observe the rules of history or, worse, conforms to the idea of a timeless time. Yet the relationships between the present, the idealized past, and the utopian future are not isomorphic in all societies. Some societies locate their visions of the good and the ideal in the past because the past they see as open-ended and renewable; some others explain the present in terms of the idiom of the past; the past for them is an allegory. Still others go back to the past to bypass cultural defeat in the present. The modern West operates on the basis of a culture to which the past is a researchable reality, close-ended and independent of the present. That is certainly not true of many other cultures which explicitly believe in the convertibility of time and space.[23] The past to them is—true to Voltaire's concept of history—a consensual fable, waiting to be interpreted and reinterpreted as an alternative in the future. It is for us to speak to it and to be spoken to in return.

[22] Forest Lin, 'Utopias East and West: The Relationships between Ancient and Modern Chinese Ideals', *Alternative Futures*, 1980, 3(3), pp. 15-31; see p. 26.
[23] See Nandy, *The Intimate Enemy*, for a discussion of this convertibility.

Towards a Third World Utopia

> Alas, having defeated the enemy, we have ourselves been defeated. . . . The . . . defeated have become victorious. . . . Misery appears like prosperity, and prosperity looks like misery. This our victory is twined into defeat.
>
> The Mahabharata[1]

I

Theories of salvation do not save. At best, they reshape our social consciousness. Utopias, too, being ideas about the end-products of salvation, cannot hope to do more. They, too, can only promise a sharper awareness and critique of existing cultures and institutionalized suffering—the surplus suffering which is born, not of the human condition, but of faulty social institutions and goals.

In this sense, all utopias and visions of the future are a language. Whether majestic, tame, or down-to-earth, they are an attempt to communicate with the present in terms of the myths and allegories of the future. When such visions are vindictive, they are a warning to us; when they are benign or forgiving towards the present, they can be an encouragement. Like history, which exists ultimately in the minds of the historian and his believing readers and is thus a means of communication, utopian or futurist thinking is another aspect of—and a comment upon—the existent, another means of making peace with or challenging man-made suffering in the present, another ethic apportioning responsibility for this suffering and guiding the struggle against it on the plane of contemporary consciousness.[2]

---

[1] *The Mahabharata*, Sauptik Parva: *10*; Slokas 9, 12, 13, trans. Manmatha Nath Dutt (Calcutta: Elysium, 1962), p. 20.

[2] Such utopianism is of course very different from the ones Karl Popper or

Thus, no utopia can be without an implicit or explicit theory of suffering. This is especially so in the peripheries of the world, euphemistically called the third world. The concept of the third world is not a cultural category; it is a political and economic category born of poverty, exploitation, indignity and self-contempt. The concept is inextricably linked with the efforts of a large number of people trying to survive, over generations, quasi-extreme situations.[3] A third-world utopia—the South's concept of a decent society, as Barrington Moore might call it —must recognize this basic reality.[4] To have a meaningful life in the minds of men, such a utopia must start with the issue of man-made suffering which has given the third world both its name and its uniqueness. This essay is an inter-civilizational perspective on oppression, with a less articulate psychology of survival and salvation as its appendage. It is guided by the belief that the only way the third world can transcend the sloganeering of its well-wishers is, first, by becoming a collective representation of the victims of man-made suffering everywhere in the world and in all past times; second, by internalizing or owning up the outside forces of oppression and, then, coping with them as inner vectors; and third, by recognizing the oppressed or marginalized selves of the first and the second worlds as civilizational allies in the battle against institutionalized suffering.[5]

The perspective is based on three assumptions. First, that as far as the core values are concerned, goodness and right ethics

Robert Nozick have in mind. See Karl Popper, 'Utopia and Violence', in *Conjectures and Refutations: The Growth of Scientific Knowledge* (London: Routledge and Kegan Paul, 1978), pp. 355–63; and Robert Nozick, *Anarchy, State and Utopia* (Oxford: Basil Blackwell, 1974), Part III.

[3] I have in mind the extremes Bruno Bettelheim describes in his 'Individual and Mass Behaviour in Extreme Situations' (1943), in *Surviving and Other Essays* (New York: Alfred Knopf, 1979), pp. 4–83.

[4] Barrington Moore, Jr., 'The Society Nobody Wants: A Look Beyond Marxism and Liberalism', in Kurt H. Wolff and Barrington Moore, Jr. (eds.), *The Critical Spirit: Essays in Honour of Herbert Marcuse* (Boston: Beacon, 1968), pp. 401–18.

[5] Though this is not relevant to the issues I discuss in this essay, the three processes seem to hint at the cultural–anthropological, the depth-psychological and the Christian-theological concerns with oppression respectively.

are not the monopoly of any civilization. All civilizations share some basic values and such cultural traditions as derive from man's biological self and social experience. The distinctiveness of a complex civilization lies not in the uniqueness of its values but in the gestalt which it imposes on these values and in the weights it assigns to its different values and subtraditions. So, certain traditions or cultural strains may, at a certain point of time, be recessive or dominant in a civilization, but they are never uniquely absent or exclusively present. What looks like a human potentiality which ought to be actualized in some distant future, is often only a cornered cultural strain waiting to be renewed or rediscovered.

Second, that human civilization is constantly trying to alter or expand its awareness of exploitation and oppression. Oppressions which were once outside the span of awareness are no longer so, and it is quite likely that the present awareness of suffering, too, will be found wanting and might change in the future. Who, before the socialists, had thought of class as a unit of repression? How many, before Freud, had sensed that children needed to be protected against their own parents? How many believed, before Gandhi's rebirth after the environmental crisis in the West, that modern technology, the supposed liberator of man, had become his most powerful oppressor? Our limited ethical sensitivity is not a proof of human hypocrisy; it is mostly a product of our limited cognition of the human situation. Oppression is ultimately a matter of definition, and its perception is the product of a worldview. Change the worldview, and what once seemed natural and legitimate becomes an instance of cruelty and sadism.

Third, that imperfect societies produce imperfect remedies of their imperfections. Theories of salvation are always soiled by the spatial and temporal roots of the theorists. Since the solutions are products of the same social experiences that produce the problems, they cannot but be informed by the same consciousness and, if you allow a psychologism, unconsciousness. Marx wrote about the process of declassing oneself and about breaking through the barriers of ideology and false consciousness; Freud, about the possibility of working through one's personal history or, rather, the defences against such history. I like to believe that these intellectual folk heroes of our times were only reflecting an analytic attitude that allows a human aggregate to work through its own past, and to critically accept, reject or use that past as a part of the aggregate's living tradition. Contrary to what they themselves believed, our heroes were reflecting a continuity with the tradition of exegesis-as-criticism that was associated with some mythopoeic traditions as well as with some forms of classical scholasticism. It is perhaps in human nature to try to design—even if with only limited success—a future unfettered by the past and, yet, inevitably informed with the past.[6]

II

What resistance does a culture face in working through its remembered past and through the limits that past sets on its worldview? What are the psychological techniques through which the future is controlled or pre-empted by an unjust system or by the experience of injustice? What are the inner checks that a society or civilization erects against eliminating man-made suffering? What can liberation from oppression in the most utopian sense mean? What is minimum freedom and what is maximum?

We cannot even begin to answer these questions without recognizing three processes which give structured oppression its resilience.

The first is the anti-psychologism which oppression breeds and from which it seeks legitimacy. The fear of soft answers to hard questions is a fear of cultures which refuse to give an absolute value to hardness itself. Many years ago Theodor Adorno and his associates had found a link between authoritar-

[6] I have argued elsewhere that whereas the modern West has specialized in speaking the language of discontinuity or creative breaks, at least some traditional societies have chosen to speak the language of continuity or renewal. See Ashis Nandy, *The Intimate Enemy: Loss and Recovery of Self under Colonialism* (New Delhi: Oxford University Press, 1983).

ian predisposition and anti-psychologism (which they, following Henry Murray, called anti-intraceptiveness).[7] Implicit in
that early empirical study of authoritarianism was the recognition that one of the ways an oppressive social system can be
given some permanence is by promoting a tough-mindedness
which considers all attempts to look within to the sources of
one's consciousness, and all attempts to grant any autonomy to
culture or mind, as something compromising, soft-headed and
emasculating. Twenty-five years afterwards Adorno recast that
argument in broader cultural terms:

Among the motifs of cultural criticism one of the most long-established
and central is that of the lie: that culture creates the illusion of a
society worthy of man which does not exist; that it concedes the
material conditions upon which all human works rise, and that,
comforting and lulling, it serves to keep alive the bad economic
determination of existence. This is the notion of culture as ideology.
. . . But precisely this notion, like all expostulation about lies, has a
suspicious tendency to become itself ideology. . . . Inexorably, the
thought of money and all its attendant conflicts extend into the
most tender erotic, the most sublime spiritual relationships. With the
logic of coherence and the pathos of truth, cultural criticism could
therefore demand that relationships be entirely reduced to their material origin. . . . But to act radically in accordance with this principle
would be to extirpate, with the false, all that was true also, all that
however impotently strives to escape the confines of universal practice, every chimerical anticipation of a nobler condition, and so to
bring about directly the barbarism that culture is reproached for
furthering indirectly. . . . Apart from this, emphasis on the material
element, as against the spirit as a lie, has given rise to a kind of
dubious affinity with that political economy which is subjected to an
immanent criticism, comparable with the complicity between police
and underworld. Since utopia was set aside and the unity of theory
and practice demanded, we have become all too practical. . . . Today
there is growing resemblance between the business mentality and
sober critical judgement.[8]

In a peculiar reversal of roles, the vulgar materialism Adorno

[7] T. W. Adorno, Else Frenkel-Brunswik, Daniel J. Levinson and R. Nevitt
Sanford, *The Authoritarian Personality* (New York: Harper, 1950).

[8] T. W. Adorno, *Minima Moralia*, trans. E. F. N. Jephcott (London: NLB,
1977), pp. 43–4.

describes is now an ally of the global structure of oppression. It colludes with ethnocide because culture to it is only an epi-phenomenon. In the name of shifting the debate to the real world, it reduces all choice to those available within a single culture, the culture affiliated to the dominant global system. In such a world, ruled by a structure that has co-opted its mani-fest critics, the search for freedom may have to begin in the minds of men, with a defiance of those cultural themes which endorse oppression by themselves endorsing the conventional defiance of oppression. As we know, oppression to be known as oppression must be felt to be so, if not by the oppressors and the oppressed, at least by some social analyst somewhere.

There is a second issue involved here. Theories of liberation built on ultra-materialism invariably inherit a certain extra-version. The various perspectives upon the future emerging from the women's liberation movement, from debates on the heritability of IQ and from the North–South conflicts all provide instances of how certain forms of anti-psychologism are used to avoid the analysis of deeper and long-term results of cruelty, exploitation and authoritarianism. The idea that the problem is exclusively the political position of women and not the politics of femininity as a cultural trait, the idea that racial discrimination begins and ends with the racial difference in IQ and does not involve the definition of intelligence as only pro-ductive intelligence and as a substitute for intellect, the belief that North–South differences involve only unequal exchange of material goods and not unequal exchange in theories of sal-vation themselves—these are all significant tributes to a global culture which is constantly seeking new and more legitimate means of short-changing the peripheries of the world. Yet, most debates around these issues assume that the impact of political and economic inequality is skin-deep and short-term. Remove the inequality and oppression, they say in effect, and you will have healthy individuals and healthy societies all around.

This anti-psychologism, partly a reaction to the over-psycho-logization of the age of the psychological man, is another means of belittling the long-term cultural and psychological effects of violence, poverty and injustice—effects which persist even when what is usually called political and economic oppression is removed. Continuous suffering inflicted by fellow human beings, centuries of inequity and deprivation of human dignity, generations of poverty, long experience of authoritarian political rule or imperialism, these distort the cultures and minds, especially the values and the self-concepts, of the sufferers and those involved in the manufacturing of suffering. Long-term suffering also generally means the establishment of powerful justifications for the suffering in the minds of both the oppressors and the oppressed. All the useful modes of social adaptation, creative dissent, techniques of survival, and conceptions of the future transmitted from generation to generation are deeply influenced by the way in which large groups of human beings have lived and died, and have been forced to live and forced to die. It is thus that institutionalized suffering acquires its self-perpetuating quality.

In sum, no vision of the future can ignore that institutional suffering touches the deepest core of human beings, and that societies must work through the culture and psychology of such suffering, in addition to its politics and economics. This awareness comes painfully, and each society in each period of history builds powerful inner defences against it. Perhaps it is in human nature to try to vest responsibility for unexplained suffering in outside forces—in fate, in history or, for that matter, in an objective science of nature or society. When successful, such an effort concretizes and exteriorizes evil and makes it psychologically more manageable. When unsuccessful, it at least keeps questions open. Predictably, every other decade we have a new controversy on nature versus nurture, a new incarnation of what is presently called sociobiology, and a new biological interpretation of schizophrenia. Biology and genetics exteriorize; psychology owns up.

The second process is a certain continuity between the victors and the victims. Though some awareness of this continuity has been a part of our consciousness for many *centuries, it is in this* century—thanks primarily to the political technology developed by Gandhi and the cultural criticisms ventured by at least some Marxist thinkers and some interpreters of Freud—that this awareness has become something more than a pious slogan. Though all religions stress the cultural and moral degradation of the oppressor and the dangers of privilege and dominance, it is on the basis of these three eponymous strands of consciousness that a major part of our awareness of the subtler and more invidious forms of oppression (which make the victims willing participants and supporters of an oppressive system) has been built. The most detailed treatment of the theme can be found in Freudian metapsychology. It presumes a faulty society which perpetuates its repression through a repressive system of socialization at an early age. Its prototypical victim is one who, while trying to live an ordinary 'normal' life, gives meaning and value to his victimhood in terms of the norms of an unjust culture. Almost unwillingly Freud develops a philosophy of the person which sees the victim as willingly carrying within him his oppressors.

In other words, Freud takes repression seriously. He does not consider human nature a fully open system which can easily wipe out the scars of man-made suffering and can thus effortlessly transcend its past. Like all history, the history of oppression has to be worked through. This piercing of collective defences is necessary, Freud could be made to say, because human groups can develop exploitative systems within which the psychologically deformed oppressors and their psychologically deformed victims (both seeking secondary psychological gains) find a meaningful life-style and mutually potentiating cross-motivations. Such cross-motivations explain the frequent human inability to be free even when unfettered, a tendency which Erich Fromm, as early as in the 1940s, called the fear of freedom.

That is the warning contained in Bruno Bettelheim's and Victor Frankl's psychoanalytic accounts of the Nazi extermination camps based on their personal experiences.[9] Both describe

[9] Bettelheim, 'Individual and Mass Behaviour'; and Victor E. Frankl, *Man's*

how some of the victims internalized the norms of the camps and became the exaggerated, pathetic, but dangerous, versions of their oppressors. Losing touch with reality out of the fear of inescapable death and trying to hold together a collapsing world, they internalized the norms and worldview of their oppressors and willingly collaborated with them, thus giving some semblance of meaning to their meaningless victimhood, suffering and death, and to the degradation and satanism of their tormentors. Elsewhere Bettelheim affirms that this was, everything said, an instance of the death drive wiping out the victim's will to live.[10] It is possible to view it also as part of a dialectic which offsets the ego defence called 'identification with the aggressor' against the moral majesty of the human spirit which, when faced with the very worst in organized repression, would rather give up the last vestiges of self-esteem and see itself as an object of deserved suffering than believe that another social group could deliberately inflict suffering without any perceivable concern for justice.[11] The killers in this case of course skilfully built upon this resilience of the victim's social self, particularly the persistence of his moral universe, and used it as a vital element in their industry of suffering.[12] The Nazis, one is constrained to admit, knew a thing or two about organized violence.

Search for Meaning (New York: Pocket Book, 1959). See also the excellent summary of related studies by Barrington Moore, Jr., *Injustice: The Social Bases of Obedience and Revolt* (New York: Macmillan, 1978), pp. 64–77. Moore also covers the untouchables of India from this point of view, pp. 55–64.

[10] Bettelheim, 'The Holocaust—One Generation Later', in *Surviving*, pp. 84–104.

[11] That this is not merely wishful thinking is partly evidenced by Helen Fein, *Accounting for Genocide: National Responses and Jewish Victimization during the Holocaust* (New York: Free Press, 1979), chapter 12. Gerda Klein says so movingly, 'Why? Why did we walk like meek sheep to the slaughter house? Why did we not fight back?... I know why. Because we had faith in humanity. Because we did not really think that human beings were capable of committing such crimes.' *All But My Life* (New York: Hill and Wang, 1957), p. 89, quoted in Terence Des Pres, *The Survivor* (New York: Oxford University Press, 1976), p. 83.

[12] Adorno, *Minima Moralia*, p. 108.

The third process which limits human visions of the future is the refusal to take full measure of the violence which an oppressive system does to the humanity and to the way of life of the oppressors. Aimé Cesaire says about colonialism that it 'works to *decivilize* the colonizer, to *brutalize* him in the true sense of the word'.[13] And, that decivilization and that brutalization one day come home to roost: 'no one colonizes innocently, . . . no one colonizes with impunity either'.[14] If this sounds like the voice of a Black Cassandra speaking of cruelties which takes place only outside the civilized world, there is the final lesson Bettelheim derives from his study of the European holocaust: 'So it happened as it must: those beholden to the death drive destroy also themselves.'[15] Admittedly we are here close to the palliatives promoted by organized religions, but even in their vulgarized forms religions do maintain some touch with the eternal verities of human nature. At least some of the major faiths have never faltered in their belief that oppressors are the ultimate victims of their own systems of violence; that they are the ones whose dehumanization goes farthest even by the conventional standards of everyday religion and everyday morality. We have come here full circle in post-modern, post-evolutionary, social consciousness. It is now fairly obvious that no theory of liberation can be morally acceptable unless it admits the continuities between its heroes and its villains and perhaps even its chroniclers.

This general continuity between the slaves and the masters apart, there are, however, the more easily identifiable secondary victims: the human instruments of violence and oppression. Their brutalization is planned and institutionalized;[16] so is the displaced hostility they attract as a 'legitimately' violent sector

[13] Aimé Césaire, *Discourse on Colonialism*, trans. Joan Pinkham (New York and London: Monthly Review Press, 1972), p. 11. Italics in the original.

[14] Ibid., p. 170.

[15] Bettelheim, 'The Holocaust', p. 101.

[16] See, for example, Chaim F. Shatan, 'Bogus Manhood, Bogus Honor: Surrender and Transfiguration in the United States Marine Corps', *Psychoanalytic Review*, 1977, *66*, pp. 585–610.

protecting those more central to the system. The ranks of the army and the police in all countries come from the relatively poor, powerless or low-status sectors of society. Almost invariably, imperfect societies arrive at a system under which the lower rungs of the army and the police are some of the few channels of mobility open to the plebians. That is, the prize of a better life is dangled before the deprived socio-economic groups to encourage them to willingly socialize themselves into a violent, empty life-style. In the process, a machine of oppression is built; it has not only its open targets but also its dehumanized cogs. These cogs only seemingly opt for what Herbert Marcuse calls 'voluntary servitude': mostly they have no escape.

Though India is one country which was colonized and ruled with the help of Indians—under colonialism the number of white men in India rarely exceeded 50,000 in a population of about 300 million—I shall pick my example of this other oppression from another society in more recent times. The American experience with the Vietnam war shows that even anti-militarism, in the form of draft-dodging or other forms of collective protest, can become a matter of social discrimination. Pacifism can be classy. The better-placed dodge more skilfully the dirty world of military violence. In the case of Vietnam, this doubly ensured that most of those who went to fight were the socially under-privileged—men who were already hurt, bitter and cynical. As is well known, a disproportionately large number of them were blacks, who neither had any respite from the system nor from their progressive, privileged fellow citizens protesting the war and feeling self-righteous. They were people who had seen and known violence and discrimination—manifest as well as latent, direct as well as institutional, pseudo-legitimate as well as openly illegitimate. Small wonder, then, that in Vietnam many of them tried to give meaning to possible death and injury by developing a pathological overconcern with avenging the suffering of their compatriots or 'buddies', by stereotyping the Vietnamese or the 'commies', or by being aggressive nationalists. The Vietnam war on this plane was a story of one set of victims setting upon another, on behalf of a reified, impersonal system of violence.[17]

III

An insight into these processes helps us visualize utopias different from the ones yielded by a straight interpretation of some of the third-world cultures. This does not mean that cultural patterns or cosmologies are unimportant. It means that the experience of man-made suffering is a great teacher. Those who maintain, or try to maintain, their humanity in the face of such experience perhaps develop the skill to give special meaning to the fundamental contradictions and schisms in the human condition—such as the sanctity of life in the presence of omnipresent death; the legitimate biological differences (between the male and the female, and between the adult, the child and the elderly) which become stratificatory principles through the pseudo-legitimate emphases on productivity, performance and potency; and the search for spirituality and religious sentiment, for human values in general, in a world where the search almost always ends up as a new sanction for the infliction of new forms of suffering. Like Marx's 'hideous heathen god who refuses to drink nectar except from the skulls of murdered men', human conciousness has used the experience of oppression to sharpen its sensitivities and see meanings which are otherwise lost in the limbo of over-socialized thinking.

One element in their vision which many major civilizations in the third world have protected with care is the refusal to think in terms of clearly opposed, exclusive, Cartesian dichotomies. For long, this refusal has been seen as an intellectual stigmata, the final proof of the cognitive inferiority of the non-white races. (Except at those moments when the idea of holism comes home to roost in someone like General Smuts or Heinrich

---

[17] This issue has been approached from a slightly different perspective in Maurice Zeitlin, Kenneth Lutterman and James Russell, 'Death in Vietnam: Class, Poverty and Risks of War', in Ira Katznelson, Gordon Adams, Philip Brenner and Alan Wolfe (eds.), *The Politics and Society Reader* (New York: David McKay, 1974), pp. 53–68.

Himmler seeking in the idea a marker of European supremacy.)
The defensive third-world response to the issue has ranged all
the way from those who hold the West guilty of not living up
to its own values (Césaire, for one, mentions the 'barbaric re-
pudiation' by Europe of Descartes' charter of universalism:
'Reason . . . is found whole and entire in each man'[18]) to those
who repudiate the values themselves—who would like Césaire
to admit that Descartes is not the last word on the intellectual
potentials of humankind or to endorse Leopold Senghor when
he declares on behalf of the non-whites: 'I feel, therefore I am'.
Rarely has the range of responses included the non-Cartesian
reply that what was once a cultural embarrassment may have
already become a reason for hope.

Many have shed tears over the genetic or social gap between
intellect and passions in the homo sapiens—the way our moral
capacities have not kept up with our cognitive skills or our left
brain with the right. Arthur Koestler and Julian Jaynes are
only the last in a long line of thinkers to feel that in this matter
nature and evolution have let us down.[19] Perhaps what looks
like a failure of nature is after all one civilization's death wish.
Perhaps reason and morality are bound to part company dramatically in a culture in which reason has to be defined so nar-
rowly. Let us not forget that Freud had, unwittingly and by
implication, worked out a psychopathology and a name for the
'Cartesian sickness'; he called it isolation, an ego defence which
isolates reason from feelings.[20] It is a defence which turns into
an inner technology Hölderlin's maxim, 'If you have understanding and a heart, show only one. Both they will damn, if
both you show together.'[21]

It is remarkable in this context that, despite all the indignity
and oppression they have faced, many defeated cultures refuse

[18] Césaire, *Discourse*, pp. 35, 51–2.

[19] Arthur Koestler, *The Ghost in the Machine* (London: Pan, 1976), ch. 18; and
Julian Jaynes, *The Origin of Consciousness in the Breakdown of the Bicameral Mind*
(Boston: Houghton Miflin, 1976).

[20] See a fuller discussion of this subject in 'Science, Authoritarianism and
Culture' in this volume.

[21] J. C. F. Hölderlin, quoted in Adorno, *Minima Moralia*, p. 197.

to draw a clear line between the victor and the defeated, the oppressor and the oppressed, the rulers and the ruled.[22] They recognize that the gap between cognition and affect tends to get bridged outside the Cartesian world, whether the gap be conceived as an evolutionary trap or as a battle between two halves of the human brain. Drawing upon the non-dualist traditions of their religions, myths and folkways, these cultures try to set some vague, half-effective limits on the objectification of living beings and on the violence which flows from it. They try to protect the faith—increasingly lost to the modern world —that the borderlines of evil can never be clearly defined, that there is always a continuity between the aggressor and his victim, and that liberation from oppressive structures outside has at the same time to mean freedom from an oppressive part of one's own self.[23] This can be read as a near-compromise with the powerful and the victorious; it can be read as cultural resistance to the 'normal', the 'rational' and the 'sane'.

The cleansing role Frantz Fanon grants to violence in his vision of a post-colonial society sounds so alien to many Africans and Asians mainly because it is insensitive to this cultural resistance.[24] Fanon admits the internalization of the oppressor. But he calls for an exorcism in which the ghost outside has to be finally confronted in violence, for it carries the burden of the ghost within. The outer violence, Fanon suggests, is the only

[22] The post-Renaissance Western preoccupation with clean divisions or oppositions of this kind is of course a part of the central dichotomy between the subject and the object, what Ludwig Binswanger reportedly calls 'the cancer of all psychology up to now'. Charles Hampden-Turner, *Radical Man* (New York: Doubleday Anchor, 1977), p. 33. For 'psychology' in the Binswanger quotation, one must of course read 'modern Western psychology'.

[23] See, for example, an interesting cultural criticism of Hinduism by a person as humane and sensitive as Albert Schweitzer (*Hindu Thought and Its Development*, New York: Beacon, 1959) for not having a hard, concrete concept of evil. For discussions of the debate around this issue, see W. F. Goodwin, 'Mysticism and Ethics: An Examination of Radhakrishnan's Reply to Schweitzer's Critique of Indian Thought', *Ethics*, 1957, 67, pp. 25–41; and T. M. P. Mahadevan, 'Indian Ethics and Social Practice', in C. A. Moore (ed.), *Philosophy and Culture: East and West* (Honolulu: University of Hawaii, 1962) pp. 579–93.

[24] Frantz Fanon, *The Wretched of the Earth* (Harmondsworth: Penguin, 1967); and *Black Skin, White Masks* (New York: Grove Press, 1967).

means of making a painful break with a part of one's own self.

If Fanon had more confidence in his culture he would have sensed that his vision ties the victim more deeply to the culture of oppression than any collaboration can. Cultural acceptance of the major technique of oppression in our times, organized violence, cannot but further socialize the victims to the basic values of his oppressors. Once given intrinsic legitimacy, violence converts the battle between two visions of the human society into a contest for power and resources between two groups sharing the same frame of values. Perhaps if Fanon had lived longer, he would have come to admit that in his method of exorcism lies a partial answer to two vital questions about the search for liberation in our times, namely, why dictatorships of the proletariat never end and why revolutions always devour their children. Hatred, as Alan Watts reminds us at the cost of being trite, is a form of bondage, too.

In our times, no one understood better than Gandhi this stranglehold of the history of oppression on the human future. That is why for him the meek are blessed only if they are, in Rollo May's terms,[25] authentically innocent, and not pseudo-innocents living out the values of an oppressive system for secondary psychological gains. Gandhi acted as if he knew that non-synergic systems, driven by zero-sum competition and search for power, control and masculinity, forced the victims to internalize the norms of the system, so that when they displaced their exploiters, they built a system which was either an exact replica of the old one or a tragicomic version of it. Hence, his concept of non-violence and non-cooperation. It stresses that the aim of the oppressed should be, not to become a first-class citizen in the world of oppression instead of a second- or third-class one, but to build an alternative world where he can hope to win back his humanity. He thus becomes a nonplayer for the existing system—one who plays another game, refusing to be either a player or a counter-player. Perhaps this is what Erik Erikson means when he suggests that

[25] Rollo May, *Power and Innocence: A Search for the Sources of Violence* (New York: Delta, 1972).

Gandhi's theory of conflict resolution imputes an irreducible minimum humanity to the oppressors and militantly promotes the belief that this humanity could be actualized.[26]

The basic assumption here is that the dehumanized tyrant is as much a victim of his system as those tyrannized; he has to be liberated, too. The Gandhian stress on austerity and pacifism comes as much from the traditional Indian principles of renunciation and monism as from a deep-seated, early-Christian belief in the superiority of the culture of the victims and from an effort to identify with that culture both as a defiance and as a testament. All his life, Gandhi sought to free the British as much as the Indians from the clutches of imperialism; the caste Hindu as much as the untouchable from untouchability. In this respect, too, he was close to some forms of Marxism and Christianity. Father G. Gutierrez, trying to reconcile Marxism and Christianity, almost inadvertently captures the spirit of Gandhi when he says:

One loves the oppressors by liberating them from their inhuman condition as oppressors, by liberating them from themselves. But this cannot be achieved except by resolutely opting for the oppressed, i.e. by combating the oppressive classes. It must be real and effective combat, not hate.[27]

This two-tier identification with the aggressor, which Gandhi so effortlessly made, is the obverse of the identification with the victim which allows a freer expression of aggressive drive. The Gandhian vision defies the temptation to equal the oppressor in violence and to regain one's self-esteem as a competitor within the same system. The vision builds on an identification with the oppressed which excludes the phantasy of the superiority of the oppressor's life-style, so deeply embedded in the consciousness of those who claim to speak on behalf of the victims of history. The vision includes the sensitivity that even those fighting an exploitative system may internalize the norms

[26] Erik H. Erikson, *Gandhi's Truth: On the Origins of Militant Nonviolence* (New York: Norton, 1969).

[27] G. Gutierrez, *A Theology of Liberation* (New York: Orbis Books, Maryknoll, 1973), p. 276.

of the system and even when openly resisting the exploiters, even when speaking of the loneliness, mental illness or decadence of the victorious, may continue to believe that the privileged are powerful not merely economically but culturally and, thus, deserve to invite jealousy or hatred, not compassion.[28]

I have tried to understand how Gandhi's future began in the present, why he viewed the global struggle against oppression as a dialectic between inter-group and within-person conflicts, and why his utopia was, in Abraham Maslow's sense, a eupsychia.[29] For better or for worse, this is the age of false consciousness; it is the awareness of the predicament of self-awareness which has shaped this century's social thinking and helped the emergence of the psychological man. In this age, Gandhi's concept of self-realization could be seen as the most serious effort to locate within the individual and in action the subject–object dichotomy (man as the maker of history versus man as the product of history; man as a self-aware aspect of nature versus man as a product of biological evolution; the ego or reality principle versus the id or pleasure principle; praxis versus dialectic or process).[30] Such a concept of self-realization is a challenge to the post-Enlightenment split in the vision of the liberated man. For two and a half centuries—starting probably with Giovanni Vico—the modern sciences of man have worked with a basic contradiction. Reacting to various forms of pre-modern fatalism, they have sought to make man the maker of his own fate—or history—by making him an object of the modern incarnations of fate—of natural sciences, social

[28] The obverse of this is of course the oppressors' search for the 'proper' worthy opponent among the oppressed. For an analysis of such a set of categories in an oppressive culture, see Nandy, *The Intimate Enemy*, chapter I.

[29] In fact, Gandhi was clearly influenced by important strands of Indian traditions which did stress such interiorization and working through. Being an internal critic of his tradition, he therefore had to do the reverse too, namely, exteriorize the inner attempts to cope with evil as only an internal state. His work as a political activist came from that exteriorization.

[30] This formulation is derived from the somewhat casual comments made by Neil Warren in his 'Freudians and Laingians', *Encounter*, March 1978, pp. 56–63; see also Philip Rieff, *The Triumph of the Therapeutic* (New York: Harper, 1966).

history, evolutionary stages and cumulative reason. Gandhi seemed to sense that this over-correction could only be remedied by worldviews which re-emphasized man's stature as a subject, seeking a more humble participation in nature and society. These are worldviews in which man is a subject by virtue of being a 'master' of nature and society *within*. They acknowledge the continuities between the suffering outside and the suffering within, and for them the self includes the experiences of the sufferings of both the self and the non-self.[31]

Using this sensitivity, Gandhi, more than anyone else in this century, tried to actualize in politics what the more sensitive social thinkers and litterateurs had already rediscovered for the contemporary awareness—that any culture of oppression is only overtly a triad of the oppressor, the victim and the interpreter. Covertly the three roles merge. A complex set of identifications and cross-identifications makes each actor in the triad represent and incorporate the other two. This view—probably expressed in its grandest form in the ancient epic on greed, violence and self-realization, Mahabharata—is the flip side of Marx, who believed that even the cultural products thrown up by the struggle against capitalism and by the enemies of capitalism were flawed by their historical roots in an imperfect society.[32] On this plane, Gandhian praxis is the logical extension of radical social criticism, for it insists that the continuity between the victim, the oppressor and the observer must be realized in action, and that

[31] Though some Western scholars like Alan Watts would like to see such location of others in the self as a typically Eastern enterprise—e.g. Alan Watts, *Psychotherapy East and West* (New York: Ballantine, 1969), this has been occasionally a part of Western philosophical concerns, too. See, for instance, Jose Ortega y Gasset, *Meditations on Quixote* (New York: Norton, 1967). Within the Marxist tradition Georg Lukacs has argued that in the area of cognition and in the case of the proletariat at least, the subject–object dichotomy is eliminated to the extent self-knowledge includes molar knowledge of the entire society. Georg Lukacs, *History and Class-Consciousness* (London: Marlin, 1971).

[32] Many of Marx's disciples sought to place Marx outside history and culture, while he himself knew better. See 'Evaluating Utopias' in this volume, where I have briefly discussed how far any theory of salvation, secular or otherwise, can shirk the responsibility for whatever is done in its name.

one must refuse to act as if some constituents in an oppressive system were morally uncontaminated.

To sum up, a violent and oppressive society produces its own special brands of victimhood and privilege and ensures a certain continuity between the victor and the defeated, the instrument and the target, the interpreter and the interpreted. As a result, none of these categories remain pure. So even when such a culture collapses, the psychology of victimhood and privilege continues and produces a second culture which becomes, over time, only a revised edition of the first. Not to recognize this is to collaborate with violence and oppression in their subtler forms. This is what most social activism and analysis begin to do once the intellectual climate becomes hostile to manifest cruelty and expropriation.

IV

A second example of the non-dual consciousness of man-made suffering can well be the refusal of many cultures to translate the principles of biopsychological continuities, such as sex and age, into principles of social stratification. Many of the major Eastern civilizations, in spite of all their patriarchal elements, see a continuity between the masculine and the feminine, and between infancy, adulthood and old age. Perhaps this is not all a matter of 'traditional wisdom'. At least in some cases it is a reaction to the colonial culture which assumed clear breaks between the male and the female, the adult and the child, and the adult and the elderly, and then used these biological differences as the homologues of secular political stratifications. In the colonial ideology, the colonizer became the tough, courageous, openly aggressive, hyper-masculine ruler and the colonized became the sly, cowardly, passive-aggressive, womanly subject. Likewise, the culture of the colonizer became the prototype of a mature, complete, adult civilization while the colonized became the mirror of a more simple, primitive, childlike cultural state. In some cases, confronted with their own ability to subjugate complex ancient civilizations, the historically-minded colonial cultures were forced to define the colonized as the homologue of the senile and the decrepit, deservedly falling under the suzerainty—and becoming the responsibility—of more vigorous cultures.

Once again I shall invoke Gandhi, who built an articulate frame of political action to counter the models of manhood and womanhood implicit in the colonial situation in India.

British colonialism in India, drawing strength from aspects of the 'mother culture', made an explicit order out of what it felt was the major strength of the Western civilization vis-à-vis the Indian. It declared masculinity to be superior to femininity which, in turn, it saw as superior to effeminacy. It then gave a structural basis to this cultural stratarchy by emphasizing the differences between the so-called martial and non-martial races of India. The obverse of this stratarchy was a similar stratarchy in some Indian subtraditions which acquired a new cultural ascendancy in British India. One example can be the various religious reform movements which stressed kshatriyahood as the future core of post-colonial Hinduism.

There were many forms of reaction to this cultural order: some Indians desperately sought instances of hyper-masculinity in the Indian past; others accepted the order and sought to excel their rulers in martial valour. Gandhi's response was to posit two alternative sets of relationships against the imperial ideology and its native versions. In one, masculinity was seen to be at par with femininity and the two had to be transcended or synthesized for attaining a higher level of public functioning. Such 'bisexuality' or 'trans-sexuality' was seen as not merely spiritually superior both to masculinity and to femininity, as in many Indian ascetic traditions, but also politically more creative. Gandhi's second model saw masculinity as inferior to femininity which, in turn, was seen as inferior to femininity in man. Here the assumption was that femininity in men, especially in the form of maternity, provided a self-critical masculinity which could subvert the values of the modern civilization more successfully than the mere affirmation of the rights of women.

I have discussed the psychological and cultural contexts of

these concepts in some detail elsewhere.[33] All I want to add here is that the formal equality which is often sought by various movements fighting for the cause of women is qualitatively different from the synergy Gandhi sought. For the former, power, achievement, productivity, work, control over social and natural resources are seen as fixed quantities on which men have held a near-monopoly and which they must now share equally with women. For Gandhi, these values are indicators of a system dominated by the masculinity principle, and the system and its values must both be seen as standing in the way of a non-oppressive world. To fight for mechanical equality, Gandhi seems to suggest, is to accept or internalize the norms of the existing system and to pay homage to masculine values under the guise of pseudo-equality.

Similarly with age. While societies which have built upon the traditions of hyper-masculinity have conceived of adulthood as the ultimate in the human life-cycle because of its productive possibilities, many of the older cultures of the world, left out of the experience of the industrial and technological revolutions, refuse to see childhood as merely a preparation for, or an inferior version of, adulthood. Nor do they see old age as a decline from full manhood or womanhood. On the contrary, each stage of life in these cultures is seen as valuable and meaningful in itself. No stage is required to derive its legitimacy from some other stage of life, nor need it be evaluated in terms of categories entirely alien to it. It has been said in recent times that alternative visions of the human future must derive their ideas of spontaneity and play from the child.[34] Implied in this very pro-

---

[33] Ashis Nandy, 'Woman Versus Womanliness: An Essay in Social and Political Psychology', *At the Edge of Psychology: Essays in Politics and Culture* (New Delhi: Oxford University Press, 1980), pp. 32–46; and *The Intimate Enemy*.

[34] See a strong plea for this in Johan Galtung, 'Visions of Desirable Societies', written for a seminar on Alternative Visions of Desirable Societies, Mexico City, 1978. This of course is complementary to the idea of 'graceful playfulness' in Ivan Illich; see his *Tools for Conviviality* (Glasgow: Fontana/Collins, 1973). For the same awareness within 'proper' Marxism see Evgeny Bogat, 'The Great Lesson of Childhood', *Eternal Man: Reflections, Dialogues, Portraits* (Moscow: Progress Publishers, 1976), pp. 288–93. The somewhat prim psychoanalytic idea of 'regression at the service of the ego' can also be viewed as an indirect plea for

posal is the tragedy of Western adulthood which has banished spontaneity and play to a small reservation called childhood, to protect the adult world from contamination. Spontaneity, play, directness of experience, and tolerance of disorder are for children or their homologues, the primitives in their sanctuaries.[35] Power, productive work and even revolutions are for mature adults and their homologues, the advanced historical societies with their experience with modern urban-industrialism and ripened revolutionary consciousness.

The dominance of the productivity principle in the modern West and the unending search for the new or the novel is a direct negation of visions which see age as giving a touch of wisdom to social consciousness and transmitting to the next generation valued elements of culture, elements which cannot be precisely formulated or transmitted as packaged products, but must be handed down to the young in the form of shared experiences. Old age is seen by the moderns primarily as a problem of management of less productive or non-productive lives. With the decline in physical prowess in men and sexual attractiveness in women, the self-image of the modern man or woman becomes something less than that of a complete human being. The pathetic worship of youth and the even more pathetic attempts to defend oneself against the inner fears of losing youthfulness and social utility—sometimes with the help of pseudo-respectful expressions such as 'senior citizens'—are produced not merely by rampant consumerism and limitless industrialism but also by a worldview associated with complex systems of oppression trying to deny the reflective or contemplative strains in the human civilization. Gerontocracy may be a false alternative to such a worldview, but it nonetheless provides

---

the acceptance of the same principle. It is possible to hazard the guess that these are all influenced in different ways by the association Christ made between childhood and the kingdom of God. That association survives within Christianity in spite of what Lloyd deMause considers to be the faith's overall thrust. See Lloyd deMause, 'The Evolution of Childhood', in *History of Childhood* (New York: Harper Torchbook, 1975), pp. 1–74.

[35] See 'Reconstructing Childhood' in this volume.

another baseline for envisioning an alternative cosmology in which age and sex would not serve as principles of social ordering, and in which respect for the qualities of old age would give completeness to youth and young adulthood, too.

V

This brings me to my third example of a non-dual vision of 'positive freedom': the cultural refusal in many parts of the savage world to see work and play as clearly demarcated modalities of human life. Many oppressed cultures, in trying to keep alive an alternative vision of a normal civilization and resisting some of the modern forms of man-made suffering, have sought to defy the modern concept of productive work and the totally instrumental concept of knowledge which goes with it. Once again I shall invoke the experience with modern colonialism, not so much because it is a shared legacy of the third world but because it did better than most other exploitative systems of the modern era in terms of having an articulate ideology, a culturally rooted legitimacy and in avoiding counterproductive violence. That colonialism was, for this very reason, one of the most dangerous forms of institutionalized violence is part of the same argument. It is not accidental that while the British empire in India lasted two centuries, the Third Reich existed for a paltry decade. Successful institutionalization of a large-scale oppressive system is not an easy achievement. It needs something more than martial skills and nihilistic passions; it needs some awareness of human limits.

One belief the colonial cultures invariably promoted was that the subject communities had a contempt for honest work, that they consisted of indolent shirkers who could not match the hard work or single-minded pursuit of productive labour of the colonizers. This was a belief sincerely held by the rulers. But sincerity in such matters, one knows, is only a defence against recognizing one's deeper need to justify a political economy which expects the subject community to work without dignity, without an awareness of being exploited, and without meaningful work goals.

The oppressed, I have argued, is never a pure victim. One part of him collaborates, compromises and adjusts; another part defies, 'non-cooperates', subverts or destroys, often in the name of collaboration and under the garb of obsequiousness. (The second part of the story creates problems for the social sciences. The modern tradition of social criticism is unidirectional. It can demystify some forms of dissent and show them to be non-dissent. It has no means of demystifying some forms of collaboration to discover secret defiance underneath. For modern social criticism equates interpretation with debunking, and this debunking must always reveal the base of 'evil' beneath the super-structure of the 'good'.) The colonized soon learnt, through that subtle communication which goes on between the rulers and the ruled, to react to and cope with the obsessive concept of productive work brought into the colonial cultures by some European and Christian subtraditions.[36] At a certain level of awareness, the subjects knew they could retaliate, tease and defy their oppressors—'fools attached to action', as the Bhagavad Gita might have called them—by refusing to share the imposed concepts of the sanctity of work and such work-values as productivity, control, predictability, discipline and utility. The differences between work and play, stressed by a repressive conscience which had to idealize colonialism as a civilizational mission, could only be resisted through an unconscious non-cooperation which included 'malingering', 'shirking' and 'indiscipline'. If this vaguely reminds the reader of the folk response of American Blacks to slavery,[37] it only shows that there is something in the experience of man-made suffering which cuts across cultures and across the folk and the classical. In India at least the much-venerated Gita was waiting

[36] On activity and work as the first postulates of a Faustian civilization, see a brief statement in Roger Garaudy, 'Faith and Liberation', in Eleonora Masini (ed.), *Visions of Desirable Societies* (London: Pergamon, 1983), pp. 47–60.

[37] Cf. E. D. Genovese, *Roll, Jordan Roll: The World the Slaves Made* (New York: Pantheon, 1974), See also Moore, *Injustice*, pp. 465–6.

to be 'misused' by those caught on the wrong side of history:

> Who dares to see action in inaction,
> and inaction in action
> he is wise, he is yogi,
> he is the man who knows what is work.[38]

This may not be the Brahmanic scholar's idea of the true meaning of the *sloka*, but what are religious texts for, if they cannot provide folk guides to survival?

If colonialism sought to do away with the human dignity of its subjects, the subjects unconsciously tried to protect their self-esteem by subverting the dignity of their rulers, by forcing the rulers to use naked force to make their subjects work, produce and be 'useful'. That is the way helpless victims are often forced to control and monitor their oppressors and to maintain an 'internal locus of control'. In their near-total impotency, they strip the authorities of the pretences to civilized authority, humane governance and, ultimately, self-respect. That is the inner logic of all domination. It ensures that if the victims are sometimes pseudo-innocent part-victims, the victors too are all too often pseudo-autonomous part-victors.

Rejection of the principle of productivity and work also means rejection of the concept of workability. Many defeated cultures have preserved with some care the banished awareness of the first and second worlds: that knowledge is valuable not because it is applicable, useful or testable, but also because it represents aesthetics, relatedness to man and nature, and self-transcendence. Certain intuitive and speculative modes of perception have come naturally to these cultures, giving rise, on the one hand, to an institutionalized dependence on music, literature, fine arts and other creative media for the expression of social thought and scientific analysis; on the other, to dependence upon highly speculative, deductive, mathematical and, even, quantitative-empirical modes of thinking as vehicles of normative passions and as expressions of religious or mystical sentiments. I have in mind here not the feeling man which

[38] *The Bhagavad Gita*, IV: 18, transcreated by P. Lal (New Delhi: Orient Paperback, 1965), p. 33.

Leopold Senghor offsets against the Cartesian man of the West, but cultures which refuse to partition cognition and affect, both as a matter of conviction and as a technique of survival.[39]

This blurring of the boundaries between science, religion and the arts is also of course a defiance of the modern concept of classification of knowledge.[40] It represents an obstreperous refusal to be converted to the modern worldview and accepting the imperialism of categories the worldview has established. It defies the total autonomy of technology and the idea of workability which has come to dominate all modern systems of knowledge. Defeated cultures know that technology now legitimates modern science and it is the spirit of instrumentality which gives a sense of personal potency, self-esteem, social status and political power in the modern sector. Technology, these cultures know, has cannibalized science.

As opposed to this culture of instrumentality, which 'works' with a concept of a universal, perfectly objective, cumulative science and admits at best only the existence of peripheral folksciences from which modern science may occasionally pick up scraps of information, the marginalized parts of the world—the second-class citizens of the third world, marginalized even in their own societies—protect their dignity by viewing the world of science as an area of a number of coexisting, universal ethnosciences, one of which has become dominant and usurped the status of the only universal and the only contemporary science.[41] Various traditional systems of medicine, artisan skills which retain the individuality of the producer and refuse to draw a line between art and craft, agricultural practices which have resisted the destructive pull of modern agronomy—these are not only aspects of a resilient cultural self-affirmation; these are indicators of a spirit which defies the power of a way of life which seeks to cannibalize all other ways of life. The third

[39] For a fuller treatment of the psychology of partitioning cognition and affect, see 'Science, Authoritarianism and Culture' in this volume.

[40] See on this subject J. P. S. Uberoi, *Science and Culture* (New Delhi: Oxford University Press, 1979).

[41] See Ashis Nandy, 'Science in Utopia: Equity, Plurality and Openness', *India International Centre Quarterly*, 1983, 10(1), pp. 47-59.

world has a vested interest in refusing to grant sanctity to a science which sees human beings and nature as the raw material for vivisectional experimentation. What seems an irrational, impractical or unworkable resistance to the products of modern science and technology in the peripheries of the world is often a deeply rooted suspicion of the instrumental vision of sectors which live off these peripheries and a desperate attempt to preserve alternative concepts of knowledge, technique and work in the interstices of the savage world.

VI

Fourth, the experience of suffering of some third world societies has added a new dimension to utopianism by sensing and resisting the oppression which comes as 'history'. By history as oppression I mean not only the limits which our past always seems to impose on our visions of the future, but also the use of a linear, progressive, cumulative, deterministic concept of history—often carved out of humanistic ideologies—to suppress alternative worldviews, alternative utopias and even alternative self-concepts. The peripheries of the world often feel that they are victimized not merely by partial, biased or ethnocentric history, but by the idea of history itself.

One can give a psychopathological interpretation of such scepticism towards history, often inextricably linked with painful, fearsome memories of man-made suffering. Defiance of history may look like a primitive denial of history and, to the extent the present is fully shaped by history in the modern perception, denial of contemporary realities. But, even from a strictly clinical point of view, there can be reasons for and creative uses of ahistoricity. What Alexander and Margarete Mitscherlich say about those with a history of inflicting suffering also applies to those who have a history of being victims:

A very considerable expenditure of psychic energy is necessary to maintain this separation of acceptable and unacceptable memories; and what is used in the defence of a self anxious to protect itself against bitter qualms of conscience and doubts about its worth is unavailable for mastering the present.[42]

The burden of history is the burden of such memories and anti-memories. Some cultures prefer to live with it and painfully excavate the anti-memories and integrate them as part of the present consciousness. Some cultures prefer to handle the same problem at the mythopoetic level. Instead of excavating for the so-called real past, they excavate for other meanings of the present, as revealed in traditions and myths about an ever-present but open past. The anti-memories at that level become less passionate and they allow greater play and lesser defensive rigidity.

What seems an ahistorical and even anti-historical attitude in many non-modern cultures is often actually an attempt on the part of these cultures to incorporate their historical experiences into their shared traditions as categories of thinking, rather than as objective chronicles of the past.[43] In these cultures, the mystical and consciousness-expanding modes are alternative pathways to experiences which in other societies are sought through a linear concept of a 'real' history. In the modern context these modes can sometimes become what Robert J. Lifton calls 'romantic totalism'—a post-Cartesian absolutism which seeks to replace history with experience.[44] But that is not a fate which is written into the origins of these modes. If the predicament is the totalism and not the romance, the *history* of civilizations after Christopher Columbus and Vasco da Gama also shows that that totalism can also come from a history which seeks to replace experience. Especially so when,

[42] Alexander Mitscherlich and Margarete Mitscherlich, 'The Inability to Mourn', in Robert J. Lifton and Eric Olson (eds.), *Explorations in Psychohistory: The Wellfleet Papers* (New York: Simon and Schuster, 1974), pp. 257–70, quotation on p. 262.

[43] See a fuller discussion of these themes with reference to Gandhi's worldview in Nandy, *The Intimate Enemy.*

[44] Robert J. Lifton, *Boundaries: Psychological Man in Revolution* (New York: Simon and Schuster, 1969), pp. 105–6. On a different plane, Alvin Gouldner has drawn attention to the close links between utopianism and ahistoricity. See his *The Dialectic of Ideology and Technology: The Origins, Grammar and Future of Ideology* (London: Macmillan, 1976), pp. 88–9.

after the advent of the idea of scientific history, history has begun to share in the near-monopoly science has already established in the area of human certitude. Albert Camus once drew a line between the makers of history and the victims of history. The job of the writer, he said, was to write about the victims. For the silent majority of the world, the makers of history also live in history and the defiance of history begins not so much with an alternative history as with the denial of history as an acreage of human certitude.

In their scepticism of history, the ahistorical cultures have an ally in certain recessive orientations to the past in the Western culture, which have re-emerged in recent decades in some forms of structuralism and psychoanalysis, in attempts to view history either as semiotics or as a 'screen memory' with its own rules of dream-work. As we well know, the dynamics of history, according to these disciplines, is not in an unalterable past moving towards an inexorable future; it is in the ways of thinking and in the choices of present times.[45]

The rejection of history to protect self-esteem and ensure survival is often a response to the structure of cognition history presumes. The more scientific a history, the more oppressive it tends to be in the experimental laboratory called the third world. It is scientific history which has allowed the idea of social intervention to be cannibalized by the ideal of social engineering at the peripheries of the world. For the moderns, history has always been the unfolding of a theory of progress, a serialized expression of a telos which, by definition, cannot be shared by communities on the lower rungs of the ladder of history. Even the histories of oppression and the historical theories of liberation postulate stages of growth which, instead of widening the victims' options, reduce them. No wonder that till now the main function of these theories has been to ensure the centrality of the cultural and intellectual experiences of a

---

[45] I need hardly add that within the modern idea of history, too, this view has survived as a latent—and, one is tempted to add, unconscious—strain. From Karl Marx to Benedetto Croce and from R. G. Collingwood to Michael Oakeshott, philosophers of history have often moved close to an approach to history which is compatible with traditional orientations to past times.

few societies, so that all dissent can be monitored and framed in the idiom of domination.

The ethnocentrism of the anthropologist can be corrected; he is segregated from his subject only socially and, some day, his subjects can talk back. The ethnocentrism towards the past mostly goes unchallenged. The dead do not rebel, nor can they speak out. So the subjecthood of the subjects of history is absolute, and the demand for a real or scientific history is the demand for a continuity between subjecthood in history and subjection in the present. The corollary to the refusal to accept the primacy of history is the refusal to chain the future to the past. This refusal is a special attitude to human potentialities, an alternative form of utopianism that has survived till now as a language alien to, and subversive of, every theory which in the name of liberation circumscribes and makes predictable the spirit of human rebelliousness.

VII

As my final example, I shall briefly discuss the so-called dependency syndrome in some third world cultures. When offset against the Occidental man's unending search for autonomy or independence, it is this syndrome in the non-Western personality which allegedly explains the origin, meaning and resilience of colonial subjugation.

Such explanations have been savagely attacked by both Césaire and Fanon as racist psychoanalysis. Césaire quotes the following words of Octave Manoni as virtually final proof of the Western psychologist's prejudice against all oppressed cultures:

It is the destiny of the Occidental to face the obligation laid down by the commandment *Thou shalt leave thy father and thy mother*. This obligation is incomprehensible to the Madagascan. At a given time in his development, every European discovers in himself the desire . . . to break the bonds of dependency, to become the equal of his father. The Madagascan, never! He does not experience rivalry with the paternal authority, 'manly protest', or Adlerian inferiority—

ordeals through which the European must pass and which are civilized forms . . . of the initiation rites by which one achieves manhood.[46]

I have not been able to locate this passage in Manoni's *Prospero and Caliban* and do not know in which context it occurs.[47] Nor do I know Manoni's politics which presumably can provide the other context of these sentences. Thus, I have to accept at face value Césaire's and Fanon's plaint that Manoni vends 'down-at-heel clichés' to justify 'absurd prejudices' and 'dresses up' the old stereotype of the Negro as an overgrown child.

Yet, I have a nagging suspicion that a third view on the subject is sustainable. That view would recognize that the modern West has not only institutionalized a concept of childhood shaped by the ideology of masculine, non-dependent adulthood and societies which represent such adulthood, it has also popularized a devastatingly sterile concept of autonomy and individualism which has increasingly atomized the Western individual. Many non-Western observers of the culture of the modern West—its life-style, literature, arts and its human sciences—have been struck by the way contractual, competitive individualism—and the utter loneliness which flows from it—dominates the Western mass society. From Friedrich Nietzsche to Karl Marx to Franz Kafka, much of Western social analysis, too, has stood witness to this cultural pathology. What once looked like independence from one's immediate authorities in the family, and defiance of the larger aggregates they represented, now looks more and more like a Hobbesian worldview gone rabid. The individual in the mass society is not only in an adversary relationship with everyone else, his individuality increasingly depends upon his becoming an independent consumption unit to which 'machines' would sell consumables and from which other machines would get work in order to produce more consumables. To the extent Manoni imputes to the Madagascan some degree of anti-individualism, to the extent the Madagascan is not a well-demarcated person, he unwittingly

[46] Octave Manoni, quoted in Césaire, *Discourse*, p. 40. Italics in the original.

[47] O. Manoni, *Prospero and Caliban: The Psychology of Colonization*, trans. Pamela Powesland, 2nd ed. (New York: Praeger, 1964).

underscores the point that modern individualism—and the insane search for absolute autonomy it has unleashed—cannot be truly separated from the thirst for colonies, *lebensraum* and domination for the sake of domination. In an interdependent world, total autonomy for one means the reduction of the autonomy of others.

Hence, while the much-maligned dependency complex may not be the best possible cultural arrangement in the face of modern oppression, it could be seen as a more promising baseline for mounting a search for more genuine social relatedness and for maturer forms of individuality than the one which now dominates the world. That baseline may not meet the exacting standards of the Westernized critics of the West in the third world, it may not yield the virile anti-Imperialism by which they swear, but those who have lived for centuries with only the extremes of relatedness and dependency will never guess that in a world taken over by the autonomy principle and by the extremes of individualism, dependency and fears of abandonment could represent a hope and a potentiality. The pathology of relatedness has already become less dangerous than the pathology of unrelatedness. What looks like an 'ego wanting in strength' in the Malagasy or a case of a 'weak ego' in the Indian can be viewed as another kind of ego strength. What looks like poor independence training in the non-achieving societies and 'willing subservience' and 'self-castration' in the Hindu may be read also as an affirmation of basic relatedness and a recognition of the need for some degree of reverence in human relations.[48] At one place in his *Discourse on Colonialism*, Césaire traces Nazism to Europe's blood-stained record in the colonies.

[48] For instance, Manoni, *Prospero and Caliban*, p. 41; G. Morris Carstairs, *The Twice Born: A Study of a Community of High-Caste Hindus* (Bombay: Allied Publishers, 1971), p. 160. I have of course in mind a galaxy of other well-motivated academics and writers, such as, to give random examples, David C. McClelland and David G. Winter, *Motivating Economic Achievement* (New York: Basic Books, 1969); Alex Inkeles and Donald H. Smith, *Becoming Modern* (London: Heinemann, 1974); Nirad C. Chaudhuri, *The Continent of Circe* (London: Chatto and Windus, 1965); V. S. Naipaul, *India: A Wounded Civilization* (London: André Deutsch, 1977); and *Among the Believers: An Islamic Journey* (London: André Deutsch, 1981).

Nazism, he says, was only a way of life coming home to roost.[49]
Césaire seems. unaware that some have already traced Nazi
satanism to the unrestrained spread in Europe, over the previous century, of the doctrines of amoral realpolitik and *sacro
egoismo* and of the 'morals of a struggle that no longer allows
for respect'.[50] All that remains to be done is to relate the colonial
impulse, too, to the search for non-reverential autonomy and
total individualism, even though the same search is now part
of many anti-colonial ideologies.

VIII

I have chosen these examples to describe what I have called
the indissoluble bond between the future of the peripheries of
the world and that of the apparently autonomous, powerful,
prosperous, imperial centres. This is necessarily an essay on
the continuity between winners and losers, seen from the losers'
point of view. The reader must have noticed that each of the
examples I have given also happens to be a live problem in
exactly those parts of the world which are commonly considered
privileged. The various forms of neo-Marxism, the various versions of the women's liberation movement, the numerous attempts to build alternative philosophies of science and technology by giving up the insane search for total control and predictability are but a recognition that the gaps between the so-called
privileged and underprivileged of the world are mostly notional.
As the peripheries of the world have been subjected to economic
degradation and political impotency and robbed of their human
dignity with the help of dionysian theories of progress, the first
and the second worlds too have sunk deeper into intellectual
provincialism, cultural decadence and moral degradation. In
my version of an old cliché, no victor can be a victor without
being a victim. In the case of nation-states as much as in the

---

[49] Césaire, *Discourse*, passim.
[50] Friedrich Meinecke and Gerhard Ritter, quoted in Renzo De Felice, *Interpretations of Fascism*, trans. Brenda H. Everett (Cambridge, Mass.: Harvard
University, 1977), pp. 17–18.

case of two-person situations, there is an indivisibility of ethical and cognitive choices. If the third world's vision of the future is handicapped by its experience of man-made suffering, the first world's future, too, is shaped by the same record.

The reader might also have noticed that I have tried to give moral and cultural content to some of the common ways in which the savage world has tried to cope with modern oppression and then projected these common ways as possibilities or opportunities. How far is this justified? After all, as one popular argument goes, history is made through the dirty process of political economy; it has no place for human subjectivity or for any defensive moralizing about human frailties and attempts to make a virtue out of necessity. Perhaps, in line with some non-modern traditions of interpretation, I could be allowed to argue that the so-called ultimate realities of political economy too could be further demystified to obtain clues to new moral visions of the human future. The frailties of human nature produced by a given social arrangement in the context of a given political economy, too, can begin to look like the baseline of a new society, once another social arrangement is envisioned. The frailty of human frailties, too, is an open question and an open text. I take heart from a brain researcher who has recently said, summing up comparative zoological work on evolution, that there also is a 'survival of the weak', and the weak do inherit the world.[51]

Such an approach is not negated by the blood-drenched history of man-made suffering in the third world—I am speaking of possibilities and opportunities, not offering a prognosis based on a trend analysis. Exactly as one cannot stop the magical mystery tours of the third world undertaken by many first-world environmentalists (in defiance of the totalist anthropocentrism and the arrogant ecocidal world-conception they see around them) by drawing attention to the poor conservationist record of the third world. For what is being proposed

[51] Paul D. MacLean, 'The Imitative–Creative Interplay of Our Three Mentalities', in Harold Harris (ed.), *Astride the Two Cultures: Arthur Koestler at 70* (New York: Random House, 1976), pp. 187–213.

is a new cultural self-expression of an ancient man–nature symbiosis, not a statistical projection of the past or the present into the future.

I hope all this will not be seen as an elaborate attempt to project the sensitivities of the third world as the future consciousness of the globe or a plea to the first world to wallow in a comforting sense of guilt. Nor does it, I hope, sound like the standard doomsday 'propheteering' which often prefaces fiery calls to a millennial revolution. All I am trying to do is to affirm that ultimately it is not a matter of synthesizing or aggregating different civilizational visions of the future. Rather, it is a matter of admitting that while each civilization must find its own authentic vision of the future and its own authenticity in future, neither is conceivable without admitting the experience of co-suffering which has now brought some of the major civilizations of the world close to each other. It is this co-suffering which makes the idea of cultural closeness something more than the chilling concept of One World which nineteenth-century European optimism popularized and promoted to the status of a dogma.[52]

The intercultural communion I am speaking about is defined by two intellectual co-ordinates. The first of them is the recognition that the 'true' values of different civilizations are not in need of synthesis. They are, in terms of basic biopsychological needs, already in reasonable harmony and capable of transcending the barriers of particularist consciousness. The principle of cultural relativism—that I write on the possibilities of a distinct eupsychia for the third world is a partial admission of such relativism—is acceptable only to the extent it accepts the

[52] As Fouad Ajami recognizes, 'The faith of those in the core in global solutions came up against the suspicions of those located elsewhere that in schemes of this kind the mighty would prevail, that they would blow away the cobwebs behind which weak societies lived. ... In a world where cultural boundaries are dismantled, we suspect we know who would come out on the top.' See Fouad Ajami, 'The Dialectics of Local and Global Culture: Islam and Other Cases', paper presented at the meeting of the group on Culture, Power and Transformation, World Order Models Project, Lisbon, 1980, mimeographed. Ajami advises us to walk an intellectual and political tightrope, avoiding both the 'pit of cultural hegemony' and 'undiluted cultural relativism'.

universalism of some core values of humankind. Anthropolog-
ism is no cure for ethnocentrism; it merely pluralizes the latter.
Absolute relativism can also become an absolute justification of
oppression in the name of ethnic tolerance, as it often becomes
in the 'apolitical' anthropologist's field report.

The second co-ordinate is the acknowledgement that the
search for authenticity of a civilization is always a search for
the other face of the civilization, either as a hope or as a warning. The search for a civilization's utopia, too, is part of this
larger quest. It needs not merely the ability to interpret and
reinterpret one's own traditions, but also the ability to involve
the often-recessive aspects of other civilizations as allies in one's
struggle for cultural self-discovery, the willingness to become
allies to other civilizations trying to discover their other faces,
and the skills to give more centrality to these new readings of
civilizations and civilizational concerns. This is the only form
of a dialogue of cultures which can transcend the flourishing
intercultural barters of our times.

Reconstructing Childhood: A Critique of the Ideology of Adulthood

There is nothing natural or inevitable about childhood. Childhood is culturally defined and created; it, too, is a matter of human choice. There are as many childhoods as there are families and cultures, and the consciousness of childhood is as much a cultural datum as patterns of child-rearing and the social role of the child. However, there are political and psychological forces which allow the concept of childhood and the perception of the child to be shared and transmitted. And it is with the political psychology of this shared concept and this transmission that I am concerned in the following analysis.

In the modern world, the politics of childhood begins with the fact that maturity, adulthood, growth and development are important values in the dominant culture of the world. They do not change colour when describing the transition from childhood to adulthood. Once we have used these concepts and linked the processes of physical and mental change to a valued state of being or becoming, we have already negatively estimated the child as an inferior version of the adult—as a lovable, spontaneous, delicate being who is also simultaneously dependent, unreliable and wilful and, thus, as a being who needs to be guided, protected and educated as a ward. Indirectly, we have also already split the child into two: his childlikeness as an aspect of childhood which is approved by the society and his childishness as an aspect of childhood which is disapproved by the society. The former is circumscribed by those aspects of childhood which 'click' with adult concepts of the child; the latter by those which are independent of the adult constructions of the child. Childlikeness is valued, sometimes even in adults. Childishness is frowned upon, sometimes even in children.

In much of the modern world the child is not seen as a homunculus, as a physically smaller version of the adult with a somewhat different set of qualities and skills. To the extent adulthood itself is valued as a symbol of completeness and as an end-product of growth or development, childhood is seen as an imperfect transitional state on the way to adulthood, normality, full socialization and humanness. This is the theory of progress as applied to the individual life-cycle. The result is the frequent use of childhood as a design of cultural and political immaturity or, it comes to the same thing, inferiority.[1] Much of the pull of the ideology of colonialism and much of the power of the idea of modernity can be traced to the evolutionary implications of the concept of the child in the Western worldview. Much of the modern awe of history and of the historical can also be traced to the same concept. Let me give two examples from the two centuries of British colonialism in India.

No better representative can be found than James Mill (1773–1836) for the sincerity of purpose which some social reformers brought into the culture of British rule in India. The nineteenth-century liberal and Utilitarian thinker's view of his private responsibility as a father meshed with his view of Britain's responsibility to the societies under its patriarchal suzerainty.[2] Mill chose to provide, almost single-handed, an intellectual framework for civilizing India under British rule. Yet he was no xenophobe. In fact, he saw the Indian empire as a training ground and an opportunity for both colonizers and colonized. Only there was a clear difference between his perceptions of the two sets of trainees. He saw Britain as the elder society guiding the young, the immature and, hence, primitive Indian society towards adulthood or maturity, and

[1] Ashis Nandy, *The Intimate Enemy: Loss and Recovery of Self under Colonialism* (New Delhi: Oxford University Press, 1983), chapter 1.

[2] For a fascinating indirect description of the interlinkages among father–son relationship, the liberal idea of progress and inter-cultural relationship, see Bruce Mazlish, *James and John Mill: Father and Son in the Nineteenth Century* (New York: Basic Books, 1975).

he felt that Indian culture required more fundamental restructuring than that required by relatively advanced Western cultures. It is thus that he provided his powerful, if indirect, ideological defence of British imperialism.

Mill's gentle civilizational mission was not the only metaphor of childhood that legitimized colonialism. Cecil Rhodes put it more clearly and, one might add, darkly: 'The native is to be treated as a child and denied franchise. We must adopt the system of despotism . . . in our relations with the barbarous of South Africa.'[3] I am unable to believe that the equation Rhodes made between childhood and barbarism was only a matter of racism. It also conveyed, I suspect, a certain terror of childhood. Rhodes was one of those persons who sensed—and had to sense—that children could be dangerous. Not merely do children define childhood, they also symbolize, once we have seen through our constructions of childhood, a persistent, living, irrepressible criticism of our 'rational', 'normal', 'adult' visions of desirable societies. Whoever does not know that 'childhood is the promise of a new world—and that new world can only be destroyed before it is born'?[4] Colonial ideology required savages to be children, but it also feared that savages could be like children.

Rudyard Kipling (1865–1936) sought to establish a relationship between the metaphor of childhood and British Imperialism on an altogether different plane. He was another one of those pathetic adults who wanted to reclaim, through his utopian vision of British rule, a lost childhood that had once been his own. Kipling, who was brought up primarily as an Indian child and whose experience in England as a child had been devastatingly cruel, spoke of the Indian as 'half savage and half child'—the former requiring civilization, the latter socialization. As I have argued elsewhere, to the extent the Indian was half child, he represented Kipling's own Indian childhood and his Indian experiences which he wanted to recover as an

[3] Cited in Chinweizu, *The West and the Rest of Us* (London: NOK, 1978), p. 403.
[4] Evgeny Bogat, 'Boys and Girls', in *Eternal Man: Reflections, Dialogues, Portraits* (Moscow: Progress Publishers, 1976), pp. 279–87; see p. 282.

adult in his heroes; to the extent the Indian was half savage, he represented Kipling's fear of his authentic Indianized self, a self he wanted to disown for the sake of his inauthentic English—and Imperialist—self, with the help of his overemphasis on laws and rules, unconditional obedience to authority, and his idea of legitimate violence inflicted or suffered for a cause.[5]

Mill and Kipling only used the growing ideological links between evolutionism and biological stratification in their culture.[6] The doctrine of progress, in the guise of models of biological and psychological development, had already promoted in post-medieval Europe, particularly in the nineteenth century, the use of the metaphor of childhood as a major justification of all exploitation. As Calvinism and the spirit of Protestantism consolidated their hold over important aspects of the European consciousness, the growth of the idea of the adult male as the ultimate in God's creation and as the this-worldly end-state for everyone was endorsed by the new salience of the productivity principle and Promethean activism, both in turn sanctified by far-reaching changes in Christianity. By about the sixteenth century the imagery of the child Christ, like that of the androgynous Christ, started becoming recessive in European Christianity. Instead, it was a patriarchal God, with a patriarchal relationship with his suffering and atoning son, that became the dominant mode in the culture. In such a culture, the child's physical weakness was already being seen as coeval with his moral and emotional weakness which needed to be corrected with the help of maturer persons. Without this correction, the child was seen to stand midway between the lower animals and humanity. In a culture in which nature, including non-human living beings, was seen as a lower stratum of God's creation, meant for man, the chosen species of God, the child as a being closer to nature was naturally considered usable—economically, socially and psychologically.

In his well-known work, Lloyd deMause faults Philippe Aries for suggesting that childhood, as we know it, is a modern

[5] See Nandy, *The Intimate Enemy.*   [6] Ibid.

creation.[7] deMause argues that children have been ill-treated throughout history, and the modern world, if anything, is somewhat kinder to the child. He is here stating an old argument which offsets modern violence against various traditional forms of institutionalized violence.

deMause is partly correct. Viewed from this side of history, the tradition of childhood is indeed the tradition of neglect, torture and infanticide. So-called parental care and education have often been a cover for the widespread social and psychological exploitation of children. Many past societies saw children as the property of their parents, sometimes without any legal protection against parental oppression. This in turn legitimized every variety of institutional violence. Mutilation of children in some societies—in the form of castration, circumcision or beautification through folk surgery—was the norm, rather than the exception. Terrorization of children for fun or for ensuring conformity was widespread. So were sodomy and other sexual abuses of children. Often these took place with the full consent not only of the victim's parents but also of society. And, above all, infanticide was not only common; it was often a way of life. It took place in the 'civilized' West and in 'pacifist' India till the middle of the nineteenth century. (Indirect female infanticide exists in many pockets of India and other traditional third-world societies even now.)

All this does suggest that mankind has progressed towards better treatment of children and that modern societies have been kinder to children than traditional societies. Such an argument, however, ignores the qualitative changes in human oppression brought about by new, impersonal, centralizing and uniformizing forces released by the modern state system, technology and, more recently, by a social consciousness dominated by mass communications. It ignores that anomic, mechanical, dispassionate, 'banal' oppression, to adapt Hannah Arendt's overworked term, is mainly a contribution of our times to the

[7] Lloyd deMause, 'The Evolution of Childhood', in Lloyd deMause (ed.), *The History of Childhood* (New York: Harper, 1975), pp. 1–76; see pp. 5–6; Philippe Aries, *Centuries of Childhood: A Social History of Family Life* (New York: Knopf, 1962).

global culture. Unlike the traditional or savage oppressor, the modern oppressor is empty within. He lives with a schizoid sense of unreality of his self and that of others. He himself is an instrument; he uses others as instruments; his reason is instrumental and he legitimizes his actions in terms of instrumentality. In sum, he lives in a world of instruments, instrumentalities and instrumentation. Such a world induces a sharp discontinuity between the oppressor and the oppressed, who no longer share the same framework of values as in the medieval witch-hunt or in pre-modern feudal land relations. They speak to each other from two sides of a soundproof glass wall. The estimated 1000 children who die every year at the hands of their parents in Britain—or the estimated casualty rate in the United States, ranging between 200,000 and 500,000 for physical abuse and between another 465,000 and 1,175,000 for severe neglect and sexual abuse—are not victims of mystification, black magic or false religious values (as in ritual child sacrifice or indirect female infanticide in India) or of poverty leading to neglect or murder.[8] They are victims of meaninglessness, the collapse of inter-generational mutuality, unlimited individualism and a system which views children as intrusions into what is increasingly considered the only legitimate dyad in the family—namely the conjugal unit. They are victims of a worldview which sees the child as an inferior, weak but usable version of the fully productive, fully performing, human being who owns the modern world.

Aries is, after all, not so specious when he speaks of childhood as a product of the post-medieval consciousness. Modern childhood did come into its own with the growth of industrialism, the spread of Protestant values, the emergence of modern technology and consolidation of colonialism. Children

[8] Richard J. Light, 'Abused and Neglected Children in America: A Study of Alternative Policies', *Harvard Educational Review*, 1973, *43*, pp. 566–7, quoted in Gilbert Y. Steiner, with the assistance of Pauline H. Milius, *The Children's Cause* (Washington, D.C.: Brookings Institute, 1976), pp. 85–9; see pp. 85–6. Steiner quotes Senator Harrison Williams as follows: 'It has become clear that brutality against children by their parents has been dramatically and tragically increasing' (p. 87).

formed one of the first social groups on which the model of the brave new world promised by these forces was tried out. For the first time in important parts of the world, normal modern adulthood could no longer be conceptualized without conceptualizing its opposite, modern childhood.

The resulting construction of childhood was not a matter of genuine false-consciousness. It did not arise from real limits to human awareness at a particular time or space. On the contrary, it involved a refusal to admit easily available data and experiences incongruent with the new ideology. If it was a false consciousness, it had built-in resistance to the recognition of its falsity.

For instance, when the Industrial Revolution gave rise to widespread use of child-labour in England, it also produced apologists of child-labour who wrote ornate, flowing prose on the good that industrial employment did to the child. Children slaving in the mills for more than twelve hours a day supposedly learned the virtues of productive work, thrift, honesty and discipline.[9] But many of these apologists also sensed that it was not incidental that the 'moral growth' of the allegedly reprobate, unsaved and savage children also helped a labour-scarce economy and produced wealth for their employers. Parents who habitually sent their children to other families to work as domestic servants and in exchange took in others' children for the same purpose, as in eighteenth-century England, mostly knew what they were doing. It was not genuine unconsciousness; it was primarily rational cost-calculations with a very thin, easily penetrable, veneer of rationalization.

I do not wish to underplay the suffering of the child in non-modern societies. Nor do I want to split hairs on the actual

---

[9] Exactly as the employers of child labourers in South Indian match and fireworks factories have recently produced elegant justifications for their employment practices in response to an indictment published by a civil rights worker. See the controversy following the publication of Smithu Kothari, 'There's Blood on Those Matchsticks: Child Labour in Shivakasi', *Economic and Political Weekly*, 2 July 1983, *18*, pp. 1191–1202; and 'Facts about Shivakasi Child Labour', *Indian Express*, 14 February 1983.

quanta of oppression involved in different societies and times. But I do doubt the glib assumptions of a theory of progress which is surreptitiously applied to life-cycles of both persons and societies. I am conscious that if the early industrial societies introduced economies of scale in the exploitation of children, most other societies, too, have tried their hand at social and emotional exploitation of their children. There is a continuity between the pre-modern and modern societies, maintained through the social inculcation in the child of culturally preferred adaptive devices such as what psychoanalysis calls the mechanisms of ego defence.[10] Though euphemistically called training in cultural values and cognitive styles, and seen as products of family socialization and organized education, there can be little doubt that many elements of such training would have been described as institutionalized brainwashing if the trainees were adults.

In spite of this continuity between the traditional and the modern ill-treatment of the child, the modern worldview is distinctive in stressing four special uses of children. Each of these uses can also be found in traditional cultures, but modern technology and communications and the spread of the values of modern life—particularly the growing instrumental view of interpersonal relationships—have given them a new reach and legitimacy.

First, there is greater sanction now for the use of the child as a projective device. The child today is a screen as well as a mirror. The older generations are allowed to project into the child their inner needs and to use him or her to work out their fantasies of self-correction and national or cultural improvement. For instance, parents may try to realize through the child their own status ambitions or to negate through him or her their own sense of economic and psychological insecurity, or they may 'bring up' the child as a double who has marital, professional and other life choices no longer open to them. Such a system can be both effective and lasting, because parents

[10] One could read this as the main thrust of the large number of culture and personality studies done in the late forties and fifties.

constitute the immediate environs of the child. The society, too, perceives them as providing a benevolent capsulating context for childhood. So when a parent acts out his or her inner conflicts on the child or tries to face the oppressions of society by using the child as a shield, he or she has the support of the entire society. Bruno Bettelheim once said that neither Hamlet's father nor King Lear had any business to impose on their progenies, on Hamlet and Cordelia respectively, the responsibility of avenging the wrongs done to the earlier generation. The parents in both cases tried to put reins on the next generation and 'saddle it with a burden of gratitude'. Hamlet's father, like Lear, 'put a private burden on his child's too weak shoulders.' And it is poetic justice, Bettelheim says, that Cordelia, willing to serve age by forgoing her right to a life of her own, suffers destruction along with her father.[11]

Unlike young adults such as Hamlet and Cordelia, younger children do not often have the option of breaking out of the social or educational 'traps' set for them. Their physical, emotional and socioeconomic vulnerability does not give them much chance of escape and they have to play out the institutional games devised for them. In many societies, by the time they gain social and economic autonomy, it is already too late for psychological autonomy;[12] they continue to carry within them the passions, hates and loves of their earliest authorities. Even when oppression becomes obvious and, thus, some subjective basis for a search for autonomy is created, as in Shakespeare's *Romeo and Juliet*, the society may turn on the young with some savagery to ensure that the search is not actualized in practice. At one time, this probably was the ultimate meaning of all blood feuds and all attempts to settle historical scores. Today, it is the source of all attempts to use children to satisfy the grandest of personal and national aspirations.

[11] B. Bettelheim, 'The Problem of Generations', in E. H. Erikson (ed.), *Youth: Change and Challenge* (New York: Basic, 1963), pp. 64–92; see pp. 69–70.

[12] This might look like an argument applicable only to modern societies, but even in traditional societies the search for psychological autonomy persists, even though more frequently in the sphere of the sacred than that of the secular.

Second, as already noted, childhood has become a major dystopia for the modern world. The fear of being childish dogs the steps of every psychologically insecure adult and of every culture which uses the metaphor of childhood to define mental illness, primitivism, abnormality, underdevelopment, non-creativity and traditionalism. Perfect adulthood, like hyper-masculinity and ultra-normality, has become the goal of most over-socialized human beings, and modern societies have begun to produce a large number of individuals whose ego-ideal includes the concept of adult maturity as defined by the dominant norms of the society. (Evgeny Bogat makes the important point that while every child is unique and while one expects that 'differences among persons and differences in character should become more apparent as people grow older . . . this is not the case. The differences sadly fade away, leaving only the memory of the wonderfully unique world of childhood.'[13]) Thus, the idea of childhood as a dystopia subtly permeates most popular myths about the lost utopia of childhood and most compensatory ideas about the beauties of childlike innocence and spontaneity. As Lloyd deMause points out in a different context, the idea of childhood as a lost utopia—found not in autobiographies but mainly in literature, myths and fantasies—is often built out of small episodes in remembered childhoods to serve as a wish-fulfilling fantasy and as a defence against traumatic childhood memories.[14] More dominant is the idea of a fearsome childhood to which one might any time regress.

Thirdly, with greater and more intense cross-cultural contacts, childhood now more frequently becomes a battleground of cultures. This is specially true of many third world societies where middle-class urban children are often handed over to the modern world to work out a compromise with cultures successfully encroaching upon the traditional life-style. For instance, even traditional rural parents may begin to send their children to modern urban schools for Western education—partly to fulfil their status ambitions and partly to create a

13 Bogat, 'Boys and Girls', p. 285.
14 deMause, 'The Evolution of Childhood'.

manageable bicultural space or an interface with the modern world within the family. Nobody who has read the lives of the reformers, political leaders and writers of nineteenth-century India can fail to notice that the Indian middle-class child became, under the growing cultural impact of British rule, the arena in which the battle for the minds of men was fought between the East and the West, the old and the new, and the intrinsic and the imposed. The autobiographies of Rabindranath Tagore and M. K. Gandhi provide excellent accounts of childhood as an area of adult experimentation in social change in mid-nineteenth-century India.[15] Both exemplify how the authors as children bore the brunt of conflicts precipitated by colonial politics, Westernized education and exogenous social institutions.

Nineteenth-century Indian childhood was not an exception. Throughout the southern world children are being made a means of reconciling the past and the present of their societies. With the accelerating pace of social change, even in many modern societies children are expected to help their elders cope with the contradictory social norms introduced into the society by large-scale technocultural changes, and to vicariously satisfy their elders' needs for achievement, power and self-esteem. (These are the needs a modern society implants in all its members but can allow only a handful to satisfy. The mythology of modernity rests on the belief that these needs can be satisfied if only an individual works hard, is adaptable and psychologically healthy. That is, there is no insurmountable institutional constraint on anyone having a sense of achievement, potency and personal worth; all failures in this respect, the modern belief goes, are actually failures of culpable individuals, not of structures.) As the modern society typically promises to meet these needs in exchange for productive, impersonal, monetized, industrial work in a competitive setting, the culture of productive work gradually takes over all other

---

[15] See Rabindranath Tagore, *Chhelebela* (Calcutta: Visvabharati University, 1944); M. K. Gandhi, *An Autobiography or The Story of My Experiments with Truth* (Ahmedabad: Navjeevan, 1927).

areas of life. It is in the modern society that we see the remarkable spectacle of even the child's early attainments in the area of sphincter control, speech, literacy and school work becoming instruments of parental drive for performance, competition, productivity and status. This is the tacit politics in the psychopathology of everyday life in many societies today.

Fourthly, societies dominated by the principle of instrumental reason and consumerism mystify the idea of childhood more than the idea of the child. This differential mystification ensures that the idea of the child is more positively cathected than the real-life child. The image of the child is in fact split and those aspects of childhood which are incongruent with the culture of adult life are defined as part of a natural savage childhood and excluded from the mythological idea of the child as a fully innocent, beautifully obedient, self-denying and non-autonomous being. In its most extreme form, the child is appreciated when he or she is least genuinely childlike or authentic—in fact, only when he or she meets the adult's concept of a good child.

The concept of a good child, derived from the objective and subjective demands of adults, finds expression in various ways. For instance, in many traditional societies such as the Indian and the Chinese, the child may be seen as a reincarnation of some familial spirit, most frequently of one's parents' parents.[16] But even when the child is seen as a good omen or as the incarnation of a good spirit, there may be a touch of instrumentality to it. Thus, a male child in a patriarchal system may be seen as a means of ensuring the continuity of lineage. He may be expected to prepare himself to look after the welfare of his ancestors, and ensure their safe passage to the life after death or look after their after-life comforts from this world through proper rituals and other religious ceremonies.

Modern parents also see children as sources of economic security, old-age insurance and as allies in the cruel world of competition, work and day-to-day politics. Many cultures and

---

[16] deMause even has a name for this; he calls it the reversal reaction. See his 'The Evolution of Childhood'.

individuals have elaborate defences against recognizing this aspect of their relationship with children. Of course, all inter-dependence is not tainted. Economic and social mutuality is no less legitimate than psychological mutuality. But when cultures help individuals to repress the contractual aspects of the adult–child relationship and help institutionalize a totally bene-volent, self-sacrificing concept of parenthood, social conscious-ness gets used to perceiving only a one-way flow of material benefits from parents to the child. The child, too, is socialized to such perceptions of benevolence and sacrifice and is con-stantly expected by the outside world as well as by his inner self to make reparative gestures towards his parents.

Thus, we seem to have come full circle to the first use of childhood we have described. If Hamlet seems too mythical a figure telling too apocryphal a story, every age has produced its version of the myth of the obligated progeny sacrificing his life to right real or imaginary wrongs done to his parents or to his parent's generation.[17]

II

Until recently, in most societies, high birth and high mortality rates ensured a plurality or near-plurality of children in the population. When the ideology of adulthood is superimposed on such social profiles, it beautifully sanctifies a subtle abridge-ment of democratic rights. Even in societies not dominated by this ideology—in societies where the child has often enjoyed a certain dignity, autonomy and, as in India, a clear touch of divinity[18]—the encroachment of the modern world on the tra-ditions of nurture and child-rearing is helping to turn the child-

---

[17] Only recently did Zulfikar Ali Bhutto, accused of political murder and condemned to death by a military junta in Pakistan, write from his death cell: 'My sons will not be my sons if they do not drink the blood of those who dare to shed my blood.' Z. A. Bhutto, *If I am Assassinated*, ed. Pran Chopra (New Delhi: Vikas, 1979).

[18] For example, on the Indian tradition of child rearing see Sudhir Kakar, *Indian Childhood: Cultural Ideals and Social Reality* (New Delhi: Oxford University Press, 1979).

hood of the third world into an ethnic variant of the first world's.

Thus, children are getting homogenized as a target as well as a metaphor of oppression and violence. Their story is becoming, to borrow Elise Boulding's expression for the history of women, another underside of history. Though some awareness of the role of the child in human civilization is reflected in religions and myths, it is mostly the lesser minds of the modern times that have emphathized with the child: an occasional Engels, not Marx, examining the political economy of the family and, indirectly, of childhood; an occasional Dickens, not Dostoevsky or Thomas Mann, anticipating twentieth-century authoritarianism in the treatment meted out to the nineteenth-century child.

It is an indicator of the power of modern consciousness that even Gandhi's Gandhism failed when it came to his own children. Though his model of social change was a majestic indictment of the metaphor of childhood legitimizing colonialism and modernity, his attempt to introduce the concept of social intervention or service in the Indian worldview did presume a non-traditional, almost Calvinist concept of the sinful, selfish child who had to be moulded into a socially useful being. Corollary-wise, in his personal life, too, Gandhi forced his sons to live in a way that would concretize his own concept of the ideal child and atone for their birth in the sin of sexuality.[19] In consequence, Gandhi's eldest son Haridas was fully destroyed. He tried his hardest not to play out Gandhi's scenario for an eldest son's life, preferring to pursue to its nadir a lifestyle defined by blind negation of his enveloping father.[20] Alcohol, prostitutes, rejection of Hinduism, and a self-centred hedonism were not only the passions of Haridas but also his flawed instruments of a self-destructive search for autonomy. In the process, he provided a classic instance of oedipal con-

[19] See on this subject Erik H. Erikson, *Gandhi's Truth* (New York: Norton, 1969); Robert Payne, *The Life and Death of Mahatma Gandhi* (New York: Dutton, 1968).

[20] Payne, *Mahatma Gandhi*.

flict in a culture which had traditionally shown a low salience
of such conflicts.

If Gandhi, too, partly gave in to the modern concept of
childhood, one can imagine the universalizing and homogeniz-
ing power of the concept. Here was a man who had not only
rejected the ideology of modernity but had also defied the implied homology between the adult–child relationship and the
West–East encounter under colonialism. Yet he was unable to
extend his dissent against the ideology of adulthood from
aggregates to persons, as he did so successfully in the case
of the man–woman relationship and the ideology of hyper-
masculinity.[21]

For intimations of that other dissent against the ideology of
adulthood one is forced to turn, paradoxically, to the best-
known ideologue of normality and adulthood of our century,
Sigmund Freud. Unlike Gandhi, Freud was totally oblivious
of the larger political use being made of the ideology of adulthood. But then he was perfectly aware of the micropolitics of
the family and that of the process of socialization.

It was Freud who first spelt out for the moderns the way
exploitation of children ensures the persistence of a tortured
childhood within each adult as a flawed consciousness. He called
such consciousness abnormal personality, one which could not
own up the remnants of an oppressed childhood within it,
because it also included norms internalized from the ideology
of adulthood. Psychopathology, in such a model, is a *double
entendre*. It is an apparently apolitical, rational attempt to cope
with inescapable memories of oppression, the so-called reality
principle being not, as social consensus and psychiatric expertise would have it, a value-neutral objectivity, but a compro-
mised apperception of reality erected by the inescapable structures of oppression. Secondly, psychopathology is a non-critical
adaptation to the pathology of a fractured interpersonal world

[21] See my 'The Final Encounter: The Politics of the Assassination of Gandhi',
in *At the Edge of Psychology* (New Delhi: Oxford University Press, 1980), pp. 70–
98.

where the unreality of conventional reality and the abnormality of conventional normality organize a child's early environment.[22]

The double meaning of psychopathology is one of Freud's major legacies. Patriarchal and conservative in personal life and overtly committed to normality and adulthood, Freud left behind in this dialectics of meaning an instrument of dissent from the ideology of normality and adulthood. It was this legacy which made the social critic Freud, in spite of all attempts to institutionalize him as a positivist applied psychologist, a reluctant political rebel and visionary of a just world. In that rebellious, 'savage' Freud, a part of the culture of modern science suspended its social-evolutionism—to affirm that childhood and adulthood were not two fixed phases of the human life-cycle (where the latter had to inescapably supplant the former) but a continuum which, while diachronically laid out on the plane of life history, was always synchronically present in each personality. And that the repression of children in the name of socialization and education was the basic model of all 'legitimate' modern repression, exactly as the ideology of adulthood (including the glorification of work, performance and productivity as normal and mature) was the prototypical theory of progress, designed to co-opt on behalf of the oppressors the visions of the future of their victims. Admittedly, the metaphor of oppression was not used by Freud, impressed as he was by the rather simple, mechanistic versions of historicism and scientism (which blunted the critical edge of his concepts such as infantile sexuality, civilization as repression, and the reality principle). But, then, all social criticism does not have the obligation to be either self-aware or self-consistent.

In sum, Freud implied (1) that the use of children for acting out the emotional conflicts of adulthood, in turn built on the ruins of an oppressed childhood, distorted the world of the

[22] In the neo-Freudian literature, the most detailed development of these themes are of course in Herbert Marcuse, *Eros and Civilization* (New York: Beacon, 1955); and R. D. Laing, *The Divided Self* (London: Tavistock, 1960). Both perspectives are compatible with a number of major traditional theories of madness.

child; (2) that 'mental illness' was only a means of protecting oneself from the inescapable arbitrary victimhood experienced in childhood; and (3) that the oppression of socialization was the root of the civilizational discontents of our times and the ultimate psychopathology of everyday life. The repression within, to use a by now worn-out expression, invariably found its social counterpart in repression without.

I suspect that the early hostility to Freud was only partly due to his concept of infantile sexuality, which in any case was implied in the Western concept of the savage, sinful child. Freud's stress on such sexuality only provided a humane interpretation of the fearful awareness that was already there in the recesses of the Western mind. The hostility to Freud was also due to his theories which hinted at the oppressiveness of the idea of adulthood and the hollowness of the theory of progress when applied to a person's life-cycle. Human childhood, Freud's metapsychology seemed to suggest, was the basic design of a society where physical and material dominance set the pace for emotional and cultural life, by forcing human subjectivity to adapt to the physical and material dependency of the child. It is the modern childhood-which-survives-childhood from which Freud sought to liberate his civilization.[23]

On this plane, Freud tried to do for the person what Gandhi tried to do for the aggregate: to free humanity from the institutionalized violence which used the metaphor of childhood and the doctrine of progress as spelt out in the dominant post-medieval concept of history. Both tacitly agreed that childhood was a culture, a quality of living and a distinctive collection of cognitive skills, emotional and motivational patterns which modernity sought to disown or repress. Liberating the child or the savage was, thus, a means of liberating the adult and the civilized from the straitjacket of 'normal' adulthood and civility.[24]

[23] Cf. Bogat, 'Boys and Girls', p. 284: 'The earlier a person leaves childhood the more infantilism there will be within him later on.' And also 'Mankind will retain its genius if it can somehow succeed in preserving the child within itself.' 'The Great Lesson or Childhood', in Bogat, *Eternal Man*, pp. 288–93, see p. 290.

[24] Nandy, *The Intimate Enemy*, chapter 1.

I doubt if any other ideological formulation could have been more subversive of the language of modernity at that point of time. The formulation sought not only to protect the child and the savage but also to alter the language of social change and to unmask the universalism of modernity as only another legitimacy for ethnocides. If we see children as carriers of a culture which is politically and socially vulnerable but is nonetheless intrinsically valuable, we also change the nature of our search for secular salvation. Analogously, cultures have a right to live not because they can be saved or promoted to a higher state of civilization but because of the alternatives they give us in their distinctive philosophies of life. Because ultimately they *are* willing to live out these alternatives on our behalf.

Freud was not a cultural relativist. In his model, childhood can be assessed in terms of the unique orientation to the natural and interpersonal worlds it represents. Cultural criticism of childhood, too, is legitimate; only it has to be ventured in the context of the biological, environmental and interpersonal demands of childhood—that is, in the context of both the psychoecology of childhood and the politics of cultures in our times. The model fears the arrogance of parents or societies which presume to 'bring up' their children; it sees family as a psychosocial space within which the culture of the adult world intersects and, sometimes, confronts the world of the child. Ideally, this sharing of space should take place on the basis of mutual respect. That it does not is a measure of our fear of losing our own selfhood through our close contacts with cultures which dare to represent our other selves, as well as a measure of our fear of the liminality between the adult and the child which many of us carry within ourselves. This is the liminality Freud worked through in his interpretation of psychopathology. This is also the liminality Gandhi had to face openly while battling the ideology of colonialism.

Liberation from the fear of childhood is also liberation from the more subtly institutionalized ethnocentrism towards past times. Elsewhere, I have discussed the absolute and total subjection of the subjects of history, who can neither rebel against the present times nor contest the present interpretations of the past. I have argued that the corollary of the modern attitude to the child is the tendency of the modern child within each modern adult to apply to past times the same doctrine of progress which is applied to the child by the adult.[25] The other name of modern childhood is personal history. Not knowing this is to be caught in the causality of history; knowing this is to reduce history to non-causal remembered past.[26] The struggle to disown one's 'childish' past in personal life is also an attempt to disown one's collective past as a pre-history or as a set of primitivisms and traditions. The struggle to own up the child within oneself is an attempt to restore wholeness in ruptured human relationships and experiences.

Is all attempt to improve or educate children, then, also an attempt to self-improve? Is every violation of children an attempt to self-destruct? Perhaps. One accepts in children what one accepts in oneself; one hates in children what one hates in oneself. Turn this into a conscious process and what looks like educating and rearing children turns out to be a pathetic attempt to compensate for unfulfilled and unrealized self-images and private ideals. Children, too, bring up their elders.

I must not end this argument leaving the impression that there is something intrinsically glamorous about childhood, or even about the innocence and victimhood of the child. That glamorization, too, is a defence against feared memories of childhood. The use of the child as a symbol of counter-cultures or utopias has often been a correlate of the use of the child as a symbol of dystopias. Children represent the contradictions and pathologies of cultures as part of an inescapable struggle for self-preservation. Adults, too, may sometimes need to be protected from them. Though the need for such protection has not arisen in the past, it may do so in the future. The vulnerability of the

<hr>

[25] 'Towards a Third World Utopia', in this volume. Also Nandy, *The Intimate Enemy*, chapter 1.

[26] Cf. Earnest Keen, 'The Past in the Future: Consciousness and Tradition', *Journal of Humanistic Psychology*, 1978, *18*, pp. 5–18.

child in the past was primarily physical. As the importance of physical power diminishes in modern social relations, the power relations between age groups may change in the same way that it is changing between the sexes. So, while the inequality between the adult and the child may not automatically decline, it may come to depend less on brute force and more on institutions, technologies and the politics of age in the future. (Witness for instance the case of the youth. The youth 'revolutions' of the late sixties in the West were also an effort to institutionalize the growing power of the youth in the Western political economy as consumers and as voters who were becoming more numerous and mobilizable both in absolute terms and in comparison with other sections of the population.)

Until now the main force behind the ill treatment of children has been the social structures and processes which have forced large sections of men and women to lose their self-esteem, and then forced them to seek that lost self-esteem through their children. From the violence-prone Spartan society (which saw its children only as future warriors for Sparta and, to test their 'toughness', exposed a large proportion of them to death at birth) to the English miners at the time of the Industrial Revolution (working fifteen to eighteen hours a day and coming back to beat up or rape their own children), violated, brutalized adulthood has been the other side of violated, brutalized children.

We thus come full circle. If violated men and women produce violated children, violated children in turn produce violated adults. Fortunately, this apparently vicious circle can be read the other way too. The ideology of adulthood has hidden the fact that children see through our hypocrisy perfectly and respond to our tolerance and respect fully.[27] Our most liberating bonds can be with our undersocialized children. And the final test of our skill to live a bicultural or multicultural existence may still be our ability to live with our children in mutuality.

[27] See an interesting indirect development of these themes in Eleonora Masini, 'Children's Images of the Future', paper presented at the meeting of World Order Models Project, Poona, India, July 1978.

A plea for the protection of children is, thus, a plea for an alternative vision of the good society on the one hand, a vision in which the plurality of cultures and paradoxically that of visions themselves are granted, and a plea for recognizing the wholeness of human personality on the other.

The Traditions of Technology

I

If the Enlightenment and the Industrial Revolution helped modern science and technology to grow, they also broke down the definitional boundaries between science and technology. The legitimacy science started earning in the seventeenth century was mostly derived from the technological achievements of the Western world. Though historians of technology are now more or less agreed that modern technology owes its growth not to science but to prior technology, the myth was gradually built that the growth of science was inextricably linked with the growth of technology and neither could be separately pursued.[1] Previously, scientists in the West did not enjoy a high social status, but taking advantage of the euphoria produced by the new technological breakthroughs—from the Elizabethan circumnavigations of the earth to the inventions reported in the evening meetings of the Royal Society, to the steam engines earnestly fumigating the European landscape— a new estate of science-and-technology emerged, which ensured its members a prestige that neither the scientist nor the technologist could earlier have dreamt of acquiring by himself. The image of the scientist as a slightly seedy natural philosopher and practitioner of an esoteric discipline, and that of the technolo-

[1] For example, D. J. Solla Price, 'Is Technology Historically Independent of Science? A Study in Statistical Historiography', *Technology and Culture*, 1965, *6*, pp. 553–68. An important attempt to provide a philosophical background to this historical difference from a non-Western perspective is in A. K. Saran, 'The Traditional Vision of Man', paper presented at the UNESCO and ICSSR Meeting of Experts on the Impact of Science and Technology on Cultural Values and the Quality of Life in Asia, Hyderabad, September 1978.

The borderline between science and technology has become more blurred in this century. I am for the moment not going into that issue. I am discussing the sources of legitimation of science in society.

gist as a humble craftsman or artisan, gradually underwent a change. Both became partners in a new, high-paying, heady enterprise called modern science. The modern scientist later was to go one better. He was to sell the idea that while each technological achievement marked the success of modern science, each technological perversity was the responsibility of either the technologist or his political and economic mentors, not that of the scientist. This is another aspect of the principle of schizoid or split legitimation I have mentioned elsewhere.[2]

Secondly, though the term 'world machine' had come to be used more and more in Western Christendom since the beginning of the thirteenth century, it was used in a favourable sense. As Lynn White reminds us, the expression used to refer not to the earth but to the skies. Things began to change in the early decades of the seventeenth century.

In 1610, . . . Galileo built an improved telescope and found spots on the sun, mountains on the moon, and satellites wheeling around Jupiter. He thus unified celestial and terrestrial physics, but instead of demechanizing the sky, he ended by mechanizing the earth. His younger contemporary, Descartes, included in this mechanization all animals; to him, and to millions who fell under his influence, they were insensible machines who, despite superficial evidences of joy or agony, had neither minds nor feelings. Only human beings had such characteristics. . . .

Then it became mankind's turn to be mechanized. . . . Darwin made man completely part of nature. There should have been nothing surprising about this. If we can believe anthropologists and the historians of non-Western cultures, most people in most times and places have thought of themselves as being such, and have found sustenance in the idea. The trouble among Westerners was that by the twentieth century the late-medieval emphasis on the importance of matter, and on mathematics as the only basis for rational certainty, had created a lifeless and impersonal image of 'nature' with which few wanted to be identified.[3]

---

[2] 'Evaluating Utopias', and 'Science, Authoritarianism and Culture', in this volume.

[3] Lynn White, Jr., 'Science and the Sense of Self: The Medieval Background of a Modern Confrontation', *Daedalus*, 1978, *107*, pp. 47–59.

It has been a long walk from Galileo to present-day protests against mechanomorphism and unlimited technology. On the way, the human experience called science has liberated Western technology from most social and cultural constraints. So much so, indeed, that the remedies against the ills of technology today are mostly sought in technology itself. For a long time, the principle 'technology for technology's sake' was seen as a harmless extension of the slogan 'science for the sake of science'. Fighting for autonomy from the medieval church and defending science against the accusations of being a runaway science, overconcerned with nature as opposed to the pure contemplation of God, the European intellectuals had no reason to suspect that modern technology would one day become the organizing principle of the world. Nor could they believe that, unlike a rabid science, a rabid technology would have an immediate impact on social relationships and the human self-image, increasingly making it impossible to think of non-technological solutions to technological problems.[4] Gradually, a means of legitimizing and thus protecting secular knowledge has become a justification for the reconceptualization and rebuilding of man himself in the image of his most successful creation.[5] Technology, an instrument, has become the model of all valued cultures and being-states.[6] The indicator of a good technology is now the extent to which it makes the more distinctive aspects

[4] Robert Sinsheimer has called this process of change in human selfhood in response to scientific and technological changes the certainty principle, the inverse of the uncertainty principle. The uncertainty principle is concerned with the impact of the observer on the observed; the certainty principle, with the expanding area of certainty and the inevitable influence of the observed on the observer. See Sinsheimer's 'The Presumptions of Science', *Daedalus*, 1978, *107*, pp. 23–35.

[5] Rolland Barthes in his 'The Brain of Einstein', *Mythologies* (New York: Hill and Wang, 1976), pp. 68–71, gives a brief but telling analysis of the ways of thinking associated with this orientation to the world. By default, B. F. Skinner and his associates provide of course an even better example of the orientation.

[6] I can find no better example of this than the 'inner speech' of behaviourism which, through the concept of operant conditioning, has reduced individualism to the principles of a biophysical machine. In the age of the psychological man, some concepts of psychological man can be very non-psychological.

of human beings redundant. A technology which displaces walking is now seen as less innovative than a technology which finds substitutes for our decision-making capacity.[7]

Thirdly, while in all societies techniques are associated with power, it was between the seventeenth and the nineteenth centuries in Western Europe that science and technology became clearly identified with the masculinity principle in the Judaeo-Christian cosmology. At one time masculinity was primarily associated with 'raw' physical power and with the refusal to depend on technological 'crutches'. This association was now gradually relegated to the mythopoetic world—to plays, novels, music and poetry. Instead, societies in the throes of the Industrial Revolution or under its indirect influence saw in their religious system what Lewis Feuer has called a 'divine sex change'.[8] Under the umbrella of Calvinism, some European cultures began to de-emphasize the imagery of the mother of the God and the feminine—rather maternal—elements in the cosmos as a part of the sacred. That feminine principle now increasingly seemed to the new generations of Europeans a part of their inglorious peasant past. Simultaneously, with the decline in the so-called magical perspectives on the natural and interpersonal worlds, there was less fear and awe of women and of their magical powers than had found expression in the great medieval witch-hunts. The European personality was now more confident in its Christianity, less fearful of its pre-Christian pagan self, and more secure in its belief that God had to be a hyper-masculine patriarch.[9] The supra-human saints and witches were now in decline and the charisma that was wrested away from nature and humanity—once, even though a mere human, one could be a 'miracle worker' as a saint or a witch—was now reinvested in the person of an omnipotent and omniscient God. In response to one's this-worldly productive work and 'masculine' self-affirmation vis-à-vis

[7] Ivan Illich seems to hint at this problem in *Tools for Conviviality* (Glasgow: Fontana/Collins, 1973), p. 15.

[8] See Lewis Feuer, *The Scientific, Intellectual and Sociological Origins of Modern Science* (New York: Basic Books, 1963).

[9] Cf. Norman Cohn, *Europe's Inner Demons* (New York: Basic Books, 1975).

nature and those close to nature, one was rewarded with a share of divine potency and science. Technology became the secular instrument of the self-redemption of the insecure male children of a heavenly Father, who retained a part of his earlier image as the supreme clock-maker, but looked less and less like a cosmic craftsman and more and more like an Imperial technocrat.

There was another cultural process at work which cut across this masculinization. It was in the seventeenth century that Europe's contacts with distant cultures and their sciences began to deepen and a more global European view of science emerged. This exposure to other sciences negated some aspects of the Christian cosmology (for instance, exposure to Chinese astronomy revealed recorded eclipses that antedated the Genesis). Already in conflict with the church, European scientists found in these experiences a handy weapon and yet, being believers, they had to reconcile the new experiences with their faith. They ended up by splitting themselves and trying to pursue faith and science separately. The result was a dramatic decline in the moral control which organized religion had earlier imposed on knowledge and techniques. Admittedly, this control had been wielded by the church in a bigoted, pettyfogging and churlish fashion, and certainly rather inefficiently. But it had provided a framework for a moral science. The new European response to science separated the spheres of morality and science, and left the latter free to define its own ethics in terms of its own needs and in terms of the secular demands of the individual and the state.[10]

II

Early post-medieval science and technology was a heady enterprise. For more than a century, there was little concerted,

[10] Cf. the position of Jürgen Habermas that 'Technocratic consciousness reflects not the sundering of an ethical situation, but the repression of "ethics" as such as a category of life.' 'Science and Technology as Ideology', in *Toward a Rational Society*, trans. J. J. Shapiro (London: Heinemann, 1971), p. 112. A different but compatible version of the same story explains the changes in European

articulate dissent in the societies which underwrote the enterprise. But when the self-doubts subsequently came, they came in a rather spectacular form. For instance, the early critics of the Industrial Revolution and the new technology in Britain—William Blake (1757–1827), William Wordsworth (1770–1850), Percy and Mary Shelley (1792–1822, 1797–1851) and Thomas Carlyle (1795–1881)—were also the most trenchant, as if they sensed that tools were only an extension of personality and the romance of new technology was only an indirect celebration of the new industrial man.[11] Some, like Wordsworth, lamenting the diminution of the communion with and the direct experience of nature, provided an indirect critique of industrialism and technologism. Others, like Marx, though never losing their innocence and awe of modern technology and giving it the status of an independent variable in social change, nevertheless granted that technology could be used to underwrite faulty social institutions in the short run.

Wordsworth's poetry could be explained away as one man's private rebellion or as the concern of a small part of the English gentry; Marx could be explained away as an irresponsible immigrant. It was more difficult to cope with Mary Shelley who expressed the inchoate fear of modern technology by turning her private nightmare into a shared myth. Surely it was not an accident that the name of her mythical maker of a monster, Frankenstein, quickly became in popular imagination also the name of the monster—a transferred epithet hinting at the latent awareness that technology could not be separated from human morality and selfhood. In other words, there went on in industrializing Europe an underground social audit of modern technology which refused to compromise with the dominant public consciousness. It was this recessive consciousness on which social critics such as Henry Thoreau (1817–

---

scientific consciousness in terms of 'a new hierarchy of knowledge by a structural inversion of the old European hierarchy of God, man and nature' in J. P. S. Uberoi, _Science and Culture_ (Delhi: Oxford University Press, 1977).

[11] A point which Bruce Mazlish has made in another way in his 'The Fourth Discontinuity', _Technology and Culture_, 1967, _8_, pp. 1–15.

1862), John Ruskin (1819–1900) and Leo Tolstoy (1828–1920) were to build, formalizing the vague discomfort with modern technology into an articulate alternative vision.

The first world war confirmed for some Western intellectuals their worst suspicions about a fully autonomous technology and their fears of reified, free-floating techniques taking over the world and shaping human destiny. Karel Capek's play introducing the robot was published in 1920, and its tone dramatically expressed Europe's anxiety about the dionysian face of modern technology. Likewise, drawing upon the Freudian vision of man, the Surrealists in André Breton's manifesto (1924) rejected not merely the core values of science and technology, but also the modern concepts of rationality and organized knowledge. The robot, they might well have argued, was not a scientific feat that had misfired; the robot was the new person who had taken over the future by being the closest approximation of the modern concept of the ideal person. The anxieties these writers and artists reflected found indirect expression in the primitivism of D. H. Lawrence (1885–1930) and the anti-therapeutics of George Bernard Shaw (1856–1950), one bemoaning the loss of man's 'natural' self, the other the loss of man's dignity and autonomy in the hand of expert technicians.[12]

There were attempts at about the same time, partly halfhearted, partly serious, to discover an alternative philosophy of technology in the works of Eastern scientists, mystics and philosophers. Men such as J. C. Bose (1858–1937), Rabindranath Tagore (1861–1941), Kakuzo Okakura (1862–1913), Aurobindo (1872–1950) and Gandhi (1869–1948) became mythopoetic figures in a section of the West. Their appeal lay only partly in their thought; the other part had to do with growing anxiety in the West about losing touch with those aspects of Western civilization the Easterners seemed to represent. The

[12] A charming expression of such fears was George Bernard Shaw's depiction of doctors and their therapeutics. See *The Doctor's Dilemma* (London: Constable, 1911). Another European response has been discussed in 'Science, Authoritarianism and Culture', in this volume.

latter looked self-confident, self-actualizing allies of the recessive Western critiques of technology. But the battle probably had already been won by modern science and technology; the critics of modern technology now more frequently had to justify their dissent by invoking the support of modern science and technology themselves.

On the other hand, even the positivists were now mounting impassioned attacks on the context, if not the content, of technology. Bertrand Russell's (1872–1970) choice of the title *Icarus* for his book on the future of science was itself revelatory. And when G. H. Hardy (1877–1947), the ultra-positivist mathematician, said that applied mathematics was 'repulsively ugly and intolerably dull' and refused to lend his mathematical skills to the British war machine, he too was participating in the widely shared, if vague, dystopia towards which technology was seemingly pushing the world.

Western scientists and consumers of technology suffered another acute bout of self-doubt after the second world war—primarily triggered by the growth of nuclear weaponry. This time even the principle of schizoid legitimation did not work. For once, the estate of science—or a section within it—began to spot its villains nearer home and did not put all the blame on imperialist powers, multinational corporations, warring capitalists and international communism. The unlucky Edward Teller, the redoubtable father of the hydrogen bomb, had not bargained for this. He probably thought that he would continue to be known for his physics and his association with nuclear weaponry would be forgotten the way the world had forgotten the discoverers of the TNT bombs, the machine gun and the gas pellets used in Auschwitz. Teller was caught on the wrong foot by the changing concept of the social responsibility of the technologist.

The post-nuclear phase of self-doubt, however, did not straightaway lead to a return to the ideas of pre-modern sciences or alternative technologies, as had happened after the first world war. Perhaps by this time the older concept of science had seeped deeper into the Western 'unconscious'. The main effort this time, summarized so well in that stale slogan 'atom for peace', was to identify an alternative techno-ethics rather than an alternative technology.

It is only since the 1970s that the autonomy and primacy of modern technology have been once again challenged by the energy crisis, by a more organized environmentalism and by a techno-ethics which dares to raise the issue of limiting technology itself. Less frequently is technology being called upon to solve the problems created by it. At a time when in some societies a growing proportion of all illnesses is iatrogenic, when the technology of weapons continues to overshadow all other technologies in financial outlays and manpower despite the stored capacity to wipe out all living beings thirty times over, and when in some societies new modes of communications and entertainment have brought down the proportion of both-way communications from about 90 per cent to less than 10 in fifty years, modern technology no longer seems sacrosanct; not even to the modern technologists. Nor does the idea of limits on technology seem an encroachment on academic freedom or progress of knowledge. In fact, the theories which vested rationality in techno-systems and technicians—and withdrew it from other human beings—lie in fragments around us, unable to cope with the fact that modern technology today has become a major bastion of irrationality and domination, apart from being a major eyesore. Our times have made it possible for a film-maker like Jean-Luc Godard to think of a cinematic climax in which the world ends neither with a bang nor with a whimper, but with a traffic jam.

III

I have recapitulated these tired details of the Western experience with modern technology to outline the cultural and psychological landscape against which one must view non-Western traditions of technology and the choices before the Asian and African societies today.

In many of these societies science traditionally enjoyed a

higher status than most forms of technology. The literati cultures in these societies often looked down upon technology as an inferior form of activity, involving low-status intervention in the real world of things and people. In Hindu society, despite the Vedic sacrificial rituals being a basic prototype of technology, the social status of the Brahmans, concerned with interpretation of the macro- and micro-cosmos of nature, was higher than the status of the artisan castes, often seen as lowly makers or manipulators of things. Theogonically, too, the Hindu goddess of learning, Saraswati, was a first-order deity, whereas the god of technology, Viswakarma, was a minor second-order deity. In the traditional cultures of Persia, Egypt and China, too, science as a part of philosophy and as a contemplative vocation seemed a natural concern or prerogative of the literati. Technology, on the other hand, though it had an important place in the cosmology, often seemed more concerned with becoming than being, overly manipulative and too obviously power-seeking.

Many traditional technologists shared this self-image. Hence, once new mobility channels opened up in their societies through modernization, they tried to impute vitalistic spirits into their machines, as if to humanize—and sanctify—their techniques and, in the process, themselves.[13] This was of course the reverse

---

[13] An excellent exploration of this theme is found in Ritwik Ghatak's film *Ajantrik*, based on a well-known short story by Subodh Ghose. In the film an old car gradually becomes an anthropomorphic reality for the hero, a lonely small-town taxi-driver. Ghatak, who started as a naïve Marxist-Leninist, does not stop with a tame formulation about commodity fetishism but sees through the rationalism and cultural evolutionism which destroys the humane vision underlying the anthropomorphism of his 'demented' hero. I have discussed how a real-life Indian scientist coped with the same issue in my 'Defiance and Conformity in Science: The World of J. C. Bose', in *Alternative Sciences: Creativity and Authenticity in Two Indian Scientists* (New Delhi: Allied Publishers, 1980), section 2.

This process is the obverse of the one Charles Chaplin explores in *Modern Times*, in which the hero himself gets increasingly mechanized as a result of his participation in a technocratic civilization. Apparently Chaplin's radicalism, being embedded in the modern culture, makes him tread along an altogether different path from Ghatak's. The problem posed by the likes of Chaplin has been discussed by Joseph Wiezenbaum, *Computer Power and Human Reason* (San Francisco: W. H. Freeman, 1976).

of the modern European experience outlined above. In the European case, nature and man were mechanized; here in traditional societies, machines, techniques or implements were anthropomorphized. Later on, thanks to the West's military, economic and political paramountcy, the former was to become an indicator of civility and progress; the latter of obscurantism or retrogression.

When towards the middle of the nineteenth century a proper theory of imperialism began to take shape, the theory used modern technology and its culture as major justifications for colonialism and its civilizing mission: Western technology was superior because Western man and the Western technological culture were better equipped for technological achievements than their savage counterparts.[14] Likewise, Western science and philosophy of science were superior to their non-Western counterparts, because Western technology was superior to the non-Western ones. In both incarnations, the theory blurred the boundary between science and technology to use them as the secular justification of Western domination. The theory of imperialism not only imputed superior objectivity and rationality to modern scientists and technologists and to societies which produced and sustained such scientists; it insisted that that objectivity and rationality should always have primacy over values such as compassion, freedom and participatory democracy.

It was this part of the theory which Western education successfully sold to the colonies. As a result, when the demands for modernization and, later on, development came from the westernized strata in the colonies, the demands included pleas to abolish traditional techniques and traditional man–nature systems and, ultimately, traditional man himself.

The recent search in parts of post-colonial Asia and Africa for a culturally-rooted technology is a reaction to this colonial sociology of technology. The growing status of traditional technology among the Westernized third-world literati is an exten-

[14] For instance Shiv Viswanathan, *Organizing for Science* (New Delhi: Oxford University Press, 1984), chapter 3.

sion of the soul-searching in the 'other' West triggered by the
crisis in the Western man–nature system both in the West and
in the third world. Weaned on Western techno-ethics, it is the
third-world elites, not the third-world societies, that have rediscovered traditional technology in recent years. For the latter,
traditional technology has always been the dominant technological mode. The first world might have to seek its alternatives
in a post-modern future; in the third world today the alternatives are often there in the present, surrounding the islands
of barricaded modernity.

A part of this interest in traditional technology is no doubt
an offshoot of what Rajni Kothari has called the new white
man's burden: the cultivated Western interest for the poor
and oppressed of the third world for whom the third world
itself, steeped in ahistoricity and oriental despotism, seems unable to care. 'Modern technology for the West and Gandhism
for the third world' could be an apt slogan for this new concern
for the non-West. But such ploys and all the duplicity involved
in the promotion of alternative technologies in the third world
cannot detract from the fact that modern Western technology
is over-organized, exploitative, regimenting and dependency-
promoting, not only in the general context of the globe, but
also in the specific context of the West.

It is, however, unlikely that the Western experience will
deter the third world from investing the modern machine with
increasingly greater charisma. The illnesses created by the
machine in the West are geographically too far away, the West's
gains from technology in terms of power and wealth are too
obvious. The third world's sufferings due to dependence and
poverty, on the other hand, seem real and immediate. Nonetheless, two interrelated processes have set the stage for a
rediscovery of traditional technology in the post-colonial
societies.

First, the rich world of traditional technical knowledge, and
the complex philosophies of science which go with it, are gradually unfolding before us as a result of the efforts of a variety
of ethno-scientists. Traditional technology now looks less of an

anachronism, an evolutionary freak in a progressive scientific estate. Previously, many traditions of technology were allowed to languish or die because of an arbitrary, hegemonistic idea of modernity.[15] Fired by the idea, many societies unleashed a process ignored by social studies of technology—the de-sanctification of entire sets of indigenous techniques. Though the old technological visions had often shown remarkable resilience and survived modernity, such secularization was sometimes taken by them as a matter of course and not even resisted. (One instance of resistance was the *charkha*, the traditional spinning wheel of India and perhaps the most remarkable example of an old technique becoming a new instrument of anti-imperialism and symbol of an alternative polity. The *charkha* had become 'obsolete' when the city-bred, Western-educated Gandhi remodelled and resurrected it as a 'traditional' technique with a new range of cultural, political and economic meanings.)

The obverse of this process can be seen in many parts of the world today where modern technology has itself been de-secularized. It has not only acquired the status of a universal culture-free technology, it has become self-justifying, transcendent and an end in itself. The human need to sacralize is evidently irrepressible; it surfaces in even the most secularized areas of life. For instance, today the specialized high technology of modern medicine and personalized transport systems can be considered traditional technologies in North America and West Europe.[16] They have three hundred years of cultural history behind them, and they have been ritually neutralized and integrated in the culture.[17] Their sanctity no longer seems any different from

[15] Ashis Nandy, 'Science in Utopia: Equality, Plurality and Openness', *India International Centre Quarterly*, 1983, *13*, pp. 47–60; also Johan Galtung, 'Towards a New International Technological Order?', in Ward Morehouse (ed.), *Science, Technology and the Social Order* (New Brunswick, N.J.: Transaction Books, 1979), pp. 270–300.

[16] Unlike the hero of *Ajantrik*, it is no longer necessary for the American car freak to anthropomorphize his car. He can attribute its magicality to its being a machine. The car has charisma for its American admirers because it is a super-machine, not because it is a super-man.

[17] Such integration involves a shift from technology-as-a-means to technology-as-an-end. It seems an attempt has been made by Hans Jonas to differentiate

the sanctity that attaches to the traditional technologies in some cultures.

Second, with the idea of pure modernity becoming increasingly unfashionable, it is becoming obvious that modern technology is only part of another tradition which has become dominant due to the skewed distribution of political and economic power in the world. It is this tradition of modernity which has cornered other scientific and technological worldviews with the help of the doctrine of progress *à la* Hobbes, Locke, Bentham and the Mills, on the one hand, and Vico, Hegel and Marx, on the other. In reality, the choice is not between traditional and modern technologies; it is between different traditions of technology, some dominant and some recessive, some endogenous and some exogenous.

This is not to deny but to stress the universal relevance of technologies. Many modern technologies, for instance, have been useful in non-modern contexts, if not as technologies at least as powerful critiques of the local traditions of technology. To the extent some of the modern measures have been fitted within traditional conceptions of health, prophylaxis and healing, the measures have enriched the traditions and given them a new resilience. Many traditional technologies have performed similar functions in the modern world. However, the tradition of modern technology has usually come as a package and developed an adversary relationship with all other traditions because of its claims to absolute, exclusive universality and an absolute right to proselytize. These claims come with three assumptions: first, there is an in-built elasticity in nature which restores the natural system after man has intervened in it; second, that technology is an ally of humanity in maintaining

---

modern from ancient technology in terms of the dialectical relationship between ends and means in the first case and a unilateral relationship between ends and means in the latter. See S. R. Carpenter, 'Developments in the Philosophy of Technology in America', *Technology and Culture*, 1979, *19*, pp. 93–9. A sophisticated statement on the ends–means question in the context of technology is by Sayyed Hussein Nasr, in *Man and Nature: The Spiritual Crisis of Modern Man* (New Delhi: Vikas, 1976); and *Western Science and Asian Cultures* (New Delhi: Indian Council of Cultural Relations, 1976). Also Saran, 'The Traditional Vision of Man'.

the centrality of the homo sapiens in the natural world; and third, that the frontiers of technology must be constantly extended and, research-wise, whatever could be done must be done. The first assumption has been negated by the scale and power of human intervention in this century; the second by the possibilities of total defeat of nature in the near future; and the third by the dangerous visions of the future ventured by many of the sciences themselves.

In the thirties, Gandhi spoke out against uncontrolled industrialization; against modern ideas of machines, productivity and consumption; against the orientation to nature identified with masculinity, civility and rationality; and against opening up every area of life to the imperial categories of modern science and technology. At the time he seemed to many progressives to be an engaging eccentric. Times have changed, and the worldview that at one time seemed so anachronistic now seems remarkably post-modern. The old or the small now looks beautiful not for its antique value or because it counters feelings of national inadequacy, but for its capacity to enlarge the area of humane technology. The rediscovery of traditional technology has come as part of a search for alternative frameworks of values or worldviews. It is an attempt to capture the magic—the legitimacy—that allows one to choose, reject or reconsider technologies in terms of criteria which are not merely technical.

This is only another way of saying that traditional technology represents a form of technology that is embedded in the cultural psychology and psycho-ecology of a community. What identifies it is not merely its technical content but the beliefs, values, symbols and myths associated with it. Assessing a technique is assessing its implicit cultural code and, ultimately, its normative presumptions.

IV

The openness to traditional technology presumes that traditional systems of knowledge in a society are not a mere compromise, serving for a time the purpose of survival, but to be replaced by modern knowledge once that purpose is served. The legitimacy of traditional technology is intertwined with that of traditional ways of life, including traditional assumptions about the desirable person and society. All cultures have ethnosciences and ethnotechniques which, in the context of those cultures, are rational and functional, if not in detail, at least in overall conception. The rationality and usefulness of specific traditional techniques must be sought within the context of this overall rationality and function—that is, within what Claude Alvares calls the algorithm of the technosystem.[18] There can be criticisms, even radical criticisms, of traditional technology but the criticisms have to be offered with some sensitivity to the cognitive map of the cultures involved. Such unromantic openness admits that a community has as much right to desanctify some of its age-old techniques as to sanctify new ones.[19]

Secondly, the concept of traditional technology has come to mean for some a shift to alternatives for reasons of capital scarcity, energy crisis, attempts to move away from an elitist, large-systems-dominated model to a decentralized political economy oriented to the needs of the poor. However, there can be a more fundamental justification for the shift. If harmony with the natural and human environment is seen as a criterion by which to judge technologies, traditional technology is neither marginal nor a 'scientific' compromise. In many societies like India, the entire modern sector produces less energy (electricity, steam, petroleum and nuclear energy taken together)

[18] Claude Alvares, *Homo Faber: Technology and Culture in India, China and the West 1500–1972* (New Delhi: Allied Publishers, 1979), chapter 1. See also Ali Baquer, Ashis Nandy, Jit Singh Uberoi, H. Y. Mohan Ram and Norman Reynolds, 'A Lesson for Science', *Mazingira*, 1979, *10*, pp. 69–74.

[19] It is a source of amusement to many modern doctors in India that chemotherapeutic and antibiotic agents are often used by practitioners of traditional medicine along with ornate traditional rituals which have nothing to do with the rituals of drug-prescribing and drug-taking as we know them in modern medicine. Yet this ritualization is one of the ways in which traditional medicine has tried to be defensively open and to bring new knowledge to accessible social groups.

than that produced by the traditional sector (coal, wood, cowdung, and animate energy) and at the cost of much higher energy inputs. In most Asian and African societies traditional technology remains the basic technological substratum, covering sometimes as much as three-fourths of the population. To relegate it to the status of a dissenting form just because modern technology is more visible in the urban-industrial sector or in the political centres of these societies is to close our eyes to the way modern technology is trying to supplant the technology and life-support systems of the people in the name of scientific progress, national security and development.

The philosophy of traditional technology affirms that every technology acquires its distinctiveness from its embeddedness in a life-style. A technosystem cannot be compared dimensionally or nomothetically with any other. The philosophy rejects the theory of progress without rejecting the idea of social criticism of technology. It refuses to see technology as a cumulative, growing system and culture as a laggard trying to adapt to technological changes *à la* William Ogburn. It insists that what looks like an advancing technology dragging an unwilling culture is often a runaway technology threatening the shared values and the life-style of a people. The idea of technological growth is, in this respect, not only an analytic category but the acceptance of a specific cultural concept of technology.

If all traditional technologies are seen as valid, competing models of universality, can one compare different traditions of technology at all or assess their relative merits or demerits? Or are they like religions, amenable more to comparative analysis than to comparative evaluation?

Absolute relativism forces us to give up all cultural criticism, for every existing techno-ethic then becomes legitimate by virtue of being only an extension of its surrounding cultural ethic. On the other hand, a nomothetic worldview stamps one set of techniques as universal and the rest as trying to approximate the universality of the first set but failing in this effort. I like to believe that the principle of cultural embeddedness merely prescribes that all comparisons among technosystems be, as some psychologists would say, morphogenic or ideographic.[20]

It follows that no ethno-technology, not even the modern one, can be historically superior to another. The view is unashamedly synchronic. Each traditional technosystem sparkles in the sun as an open but integrated product of a total life-style, flawed by the conflicts and dialectics of that life-style, yet neither comprehensible nor assessable without an understanding of the inputs the technosystem makes to that life-style. The criteria of a good techno-ethic lies in this sense outside technology. That may make technology seem less a pure product of human cognition—which in the first place it never was—and more an expression of the total human personality, which it is not often given a chance to become.

---

[20] For instance Gordon W. Allport, 'The General and the Unique in Psychological Science', *Journal of Personality*, 1962, *30*, pp. 403–22.

Science, Authoritarianism and Culture:
On the Scope and Limits of Isolation
Outside the Clinic

I

M. N. Roy was always certain that he was fighting for the modern world. He was openly anti-traditional and openly a rationalist who sought to transcend his culture. But is this the whole truth about him? Does commitment to one's culture have to be explicit and aggressive? Could it not be implicit and unconscious? When Roy as a young revolutionary, escaping from the colonial police, changed his name from Narendra Nath to Manabendra Nath, was it only carelessness that he retained in his new name the meaning of the old? Or was it a clue to his deeper awareness of the need to recognize continuities and traditions? In his later life when he used the concept of cultural renaissance, did he mean what he said or did he have only a naissance in mind?

One discipline's trivia are always another discipline's life-blood. I venture the guess that Roy here was unwittingly hinting at a psychological process which has often been inaccessible to the modern experience, namely the affirmation of traditions and cultural continuities in the face of the homogeneity that the modern world imposes in the name of universalism. Is such affirmation a pathology, a shared irrationality or a nostalgia which an old society must overcome to enter the contemporary world? I shall try to amplify here one possible answer to this question, exploring in the process one aspect of the linkage between modern science, authoritarianism and culture.

II

Every age has its prototypical violence. The violence of our age is based not so much on religious fanaticism or tribal blood feuds, as on secular, objective, dispassionate pursuit of personal and collective interests. Every age also probably has a cut-off point when the self-awareness of the age catches up with the organizing principles of the age, when for the first time the shared public consciousness begins to own up or rediscover— often through works of art or speculative thought—what the seers or the lunatics had been saying beyond the earshot of the 'sane', 'normal', 'rational' beings who dominate the public discourse of the time.

Thus, it was the mindless blood-letting of the first world war which created a new awareness of an old psychopathology of our times. As the range of human violence and the role of science in that violence began to weigh on the social conscience, a number of European intellectuals woke up at about this time to the dangerous human ability to separate ideas from feelings and to pursue ideas without being burdened by feelings. With the advantage of hindsight, one could trace the cultural sanction for this ability to changes in European cosmology in the sixteenth and seventeenth centuries. It was then that the anthropomorphic worldview began to give way to a mechanomorphic view of nature and society. It was then that what psychoanalysts may call a projective science—a science heavily dependent on the psychological capacity to project into the outer world the scientist's inner feelings and panpsychic fantasies—began to give way to a new concern with objective impersonal pictures of nature and society as the goal of knowledge and as an indicator of progress. But it was the first world war which for the first time shook the popular faith in perpetual progress through increasingly objective science. And as all other traditions of science were moribund in the West and some of them were living in the East, the war, also for the first time, led to a serious, self-conscious effort to involve the East in Europe's self-doubts.

Sigmund Freud first gave a name to this splitting of cognition and affect. He called it isolation. He described it as an ego defence, a psychological mechanism which helped the human mind to cope with unacceptable or ego-alien inner impulses and external threats. According to Freud, the individual some-times isolated an event, idea or an act by cauterizing it emotion-ally and by preventing it from becoming a part of his significant experience. The event, idea or the act was not forgotten; it was reincorporated into consciousness after being deprived of its affect.[1] This did not, Freud granted, really free ideas or actions from feelings. It merely replaced conscious associations by un-conscious ones and displaced the affect to other ideas or events. (Freud also noted the heavy use of isolation in the character disorder called obsession-compulsion. The connection, by itself, may not seem important but it acquires a different meaning if we remember that some psychological works have referred to the obsessive-compulsive associations of modern authoritarian-ism. I shall come back to this.)

Later, two second-generation psychoanalysts, Anna Freud and Otto Fenichel, were to define isolation more formally. Here is Fenichel on the subject, in his well-known textbook:

> The most important special case of this defence mechanism is the isolation of an idea from the emotional cathexis (load of feelings) that originally was connected with it . . . In discussing the most excit-ing events, the patient remains calm but may then develop at quite another point an incomprehensible emotion, without being aware of the fact that the emotion has been displaced . . .

> The normal prototype is the process of logical thinking, which actually consists of the continued elimination of affective associations in the interest of objectivity. . . . Compulsion neurotics, in their isolation activities, behave like caricatures of normal thinkers. . . . they always desire order, routine, system.[2]

Such a definition, however clinical or sterilized it may sound to its author, already verges on social criticism. It admits that

[1] Sigmund Freud, *Inhibitions, Symptoms and Anxiety* (1926), Standard Edition, Vol. 20 (London: Hogarth, 1959).

[2] Otto Fenichel, *The Psychoanalytic Theory of Neurosis* (New York: Norton, 1945), p. 156.

order, routine and system are not absolute values, that an over-commitment to them could be an illness. It also implies that objectivity, and the separation of the observer from the observed, is not an unmixed blessing; sometimes it can hide fearsome passions.

Psychoanalysis was not alone. At about the same time that the young discipline was forging the concept of isolation, the surrealist manifestos of André Breton and his associates were rejecting conventional rationality and indirectly attacking the growing use of isolation in modern life. Salvador Dali, for instance, 'absurdized' in his art and life exactly this psychopathology. His watches which melted and his machines which were part-human were but instances where the lost affect was made to re-enter social perceptions, to shock or to enchant. Some years afterwards George Orwell was scandalized when the middle-aged Dali put into his memoirs, with obvious relish, the following incident which took place when Dali was six years old:

While crossing the hall I caught sight of my little three-year-old sister crawling unobtrusively through a doorway. I stopped, hesitated a second, then gave her a terrible kick in the head, as though it had been a ball, and continued running, carried away with a 'delirious joy' induced by this savage act.[3]

Orwell correctly guessed that Dali's pathology tied up with the pathology of a period and quoted a rhyme popular around 1912 to make his point:

> Poor little Willy is crying so sore,
> A sad little boy is he,
> For he's broken his little sister's neck
> And he'll have no jam for tea.[4]

As if to prove Orwell right, Dali's autobiography became a best-seller.

---

[3] Quoted in George Orwell, 'Benefit of Clergy: Some Notes on Salvador Dali' (1944), in *Decline of the English Murder* (Harmondsworth: Penguin, 1965), pp. 20–30.

[4] From Harry Graham's *Ruthless Rhymes for Heartless Homes*, quoted in Orwell, 'Benefit of Clergy', p. 29.

Within a decade or two, a number of movements in literature and the arts caught up with the same pathology, often brilliantly though rarely self-consciously. Thus, many of the comic devices of Bertolt Brecht can be read as attempts to tear away the mask which isolation allows the industrial society to wear. When one laughs with Brecht, one also laughs at the subversion of the defence of isolation. Under the structure of isolation lies, Brecht seems to say, psychopathic hypocrisy or sheer self-deceit. Those who have seen or read his *Mr Puntilla* (1940) will know that it is the story of a businessman whose personality is split. He is a heartless calculating machine when sober; humane and lovable when drunk. When sober, pathological isolation is the main feature of his personality. When drunk, the feelings he dissociates from ideas and actions re-emerge uncensored and get reattached to his ideas and actions. That this happens only when he is drunk is, of course, Brecht's final comment on the psychopathology of modern society.

Another instance from the popular arts could be Charles Chaplin's *Monsieur Verdoux* (1947), a black comedy set against the collapse of values in inter-war Europe. The movie makes subtle use as well as criticism of the mechanism of isolation. It tells the story of a lovable psychopath who marries and then charmingly kills his wives for money. Chaplin offsets this isolation against the larger isolation taking place in the European society and against the isolation that the movie induces in viewers. As we isolate the acts of murder from the emotions they should arouse, we laugh at Chaplin's murders and sympathize with his hero, who does on a small scale what societies do on a grander scale.[5]

Chaplin's folk philosophy found its clearest expression in Orwell's essay on the use of the English language to sterilize thinking and to cover up violence and cruelty:[6]

---

[5] More recent examples of successful attempts to create black comedies on the basis of the human capacity to isolate are Stanley Kubrick's *Dr Strangelove* and *A Clockwork Orange*. Incidentally, black comedy as a genre is nearly absent in Indian and other non-modern creative traditions. It is probably a modern innovation.

[6] George Orwell, 'Politics and the English Language' (1946), in *Inside the Whale and Other Essays* (Harmondsworth: Penguin, 1957), pp. 143–57.

In our time, political speech and writing are largely the defence of the indefensible. Things like the continuance of British rule in India, the Russian purges and deportations, the dropping of the atom bombs on Japan, can indeed be defended, but only by arguments which are too brutal for most people to face. . . . Thus political language has to consist largely of euphemism, question-begging and sheer cloudy vagueness. Defenceless villages are bombarded from the air, the inhabitants driven out into the countryside, the cattle machinegunned, the huts set on fire with incendiary bullets: this is called *pacification*. Millions of peasants are robbed of their farms and set trudging along the roads with no more than they can carry: this is called *transfer of population* or *rectification of frontiers*. People are imprisoned for years without trial, or shot in the back of the neck or sent to die of scurvy in Arctic lumber camps: this is called *elimination of unreliable elements*.[7]

Orwell wrote this in the mid-forties. Around the same time, basing themselves on two major empirical studies done from Freudian and Marxist vantage grounds, some scholars began to mention the over-use of isolation by the fascist personality. Erich Fromm described the authoritarian person not only as sado-masochistic but as having a mechanical, rigid mode of thinking characterized by isolation. Fascism, he said, thrived on the objectification of persons and groups.[8] Theodor Adorno and his associates, too, wrote about the 'empty, schematic, administrative fields' in the mind of the fascist and about the constriction of his inner life.[9] The fascist, they said, partitioned his personality into more or less closed compartments. He had a narrow emotional range and he rejected emotional richness, intuitions and the softer side of life. He admired organizations and their formal hierarchies and he sought security in isolating hierarchical structures.[10]

If all this seems overly psychological, there were scholars who traced the institutional roots of European Fascism to the

[7] Ibid., p. 153.

[8] Erich Fromm, *Escape from Freedom* (New York: Holt, 1941).

[9] T. W. Adorno, Else Frenkel-Brunswick, D. Levinson and N. Sanford, *The Authoritarian Personality* (New York: Norton, 1950).

[10] All these traits were seen as aspects of the obsessive-compulsive personality of the fascist. I have already mentioned that in his earlier formulation of the problem Freud had posited a close bond between isolation and obsession-compulsion.

separation of ideas from feelings, and of the rational from the irrational. Friedrich Meinecke, for instance, located the origins of National Socialism in the ancient 'bipolarity extending throughout life of the Western Man' between the utilitarian which was stressed and the spiritual which was suppressed, to the excessive emphasis on the 'calculating intelligence', and to a Machiavellian rebirth which transformed Machiavellianism from a trait of the aristocracy to that of the middle classes and, later on, the masses.[11] Alexander and Margarete Mitscherlich's psychological profile of post-war Germany fits the pattern:

> The most important collectively practiced defense is to withdraw cathectic energies from all processes related to enthusiasm for the Third Reich, idealization of the Führer and his doctrine, and, of course, actual criminal acts. . . . The community of those who had lost their ideal 'leader', the representative of a commonly shared ego ideal, managed to avoid self-devaluation by breaking all affective bridges linking them to the immediate past. . . . Had it not been counteracted by these defense mechanisms—of denial, isolation, transformation into the opposite, and above all withdrawal of interest and affect, that is to say of rendering memories of the whole period of the Third Reich devoid of feeling—a condition of extreme melancholia would have been inevitable for a large number of people in postwar Germany . . .[12]

Hannah Arendt was to later contribute to the same awareness with her portrait of Adolf Eichmann, a plain-thinking, non-ideological, hard-working, bureaucratic killer who saw his genocidal responsibility as a problem of efficiency, organization and objective planning.[13] Arendt recognized that Eichmann was the ultimate product of the modern world, not because he established a new track-record in monstrosity but because he

[11] Friedrich Meinecke, *The German Catastrophe: Reflections and Recollections*, trans. Sidney B. Fay (Cambrige, Mass.: Harvard University, 1950), pp. 37, 51. Cited in Renzo De Felice, *Interpretations of Fascism*, trans. Brenda H. Evrett (Cambridge, Mass.: Harvard University, 1977), pp. 15–17.

[12] Alexander Mitscherlich and Margarete Mitscherlich, 'The Inability to Mourn', in Robert J. Lifton and Eric Olson (eds.) *Explorations in Psychohistory: The Wellfleet Papers* (New York: Simon and Schuster, 1974), pp. 257–70; see pp. 264, 266, 268–69.

[13] Hannah Arendt, *Eichmann in Jerusalem* (New York: Viking, 1963).

typified the evil that grew out of everyday isolation rather than from the satanism which comes from unbridled passions. (Appropriately enough, the great majority of his victims too were 'utterly unable to comprehend what had happened to them. . . . They had no consistent philosophy which could protect their integrity as human beings, which could give the strength to make a stand. . . . They had obeyed the law handed down by the ruling classes, without ever questioning its wisdom.'[14] Evidently, in Eichmann's industry of death, mechanical, bureaucratic acceptance faced a mechanical, bureaucratic death machine.)

Thus, since the 1920s, sensitive minds were warning us about the dangers of affectless sanitized cognition, about what Robert Pirsig calls 'a noncoalescence between reason and feeling'.[15] And, by the early fifties it was clear to many that fascism was the typical psychopathology of the modern world, for it merely took to logical conclusions what was central to modernity, namely the ability to partition away human cognition and pursue this cognition unbridled by emotional or moral constraints.

III

Only one area of modern life escaped the full impact of the critique of isolation: modern science. There were reasons for this. Modern science was structured isolation. The values of objectivity, rationality, value-neutrality and inter-subjectivity were definitionally the values of the modern scientific worldview. And these values *did* heavily draw upon the human capacity to isolate. Moreover, there was a latent awareness in the society that science was, at times, isolation at its best and at its most exciting, that somehow the abstractive and generaliz-

---

[14] Bruno Bettelheim, *Surviving and Other Essays* (New York: Knopf, 1979), pp. 56–7.
[15] Robert Pirsig, *Zen and the Art of Motorcycle Maintenance* (London: Corgi, 1976), p. 162.

ing capacities of science were closely related to the process of isolation. Theodore Kroeber, a relatively unknown psychologist, once perspicaciously described objectivity as a coping mechanism, which was the healthy counterpart of the defence of isolation.[16] Science as a personal search for truth and as a means of human self-realization seemed to be a form of this creative objectivity. It did not seem that isolating to many. The attacks of the artists, writers and the fashionable mystics, in contrast, were bound to wash off as eccentric responses to the creative isolation of modern science.

Moreover, a part of the attack on science was diverted to technology. As the dehumanizing and mechanomorphic aspects of technology became obvious after the first world war, there emerged the view that questions of ethics applied mainly to technology, not to science. This was certainly the argument of the major social critics who shaped the popular response to science. Take for instance the two literary figures who helped to bring us up in the first half of this century: George Bernard Shaw and H. G. Wells. Shaw wrote savage indictments of modern technology in *Major Barbara* and *The Doctor's Dilemma*. But he also wrote fiery tracts pleading for more modern scientific management of societies. Wells's science fiction could be read as a trenchant critique of a science contaminated by human greed and violence. (*The Island of Dr Moreau* and its vivisectional horrors, one may argue, were distant in only geographical terms; psychologically they were right in the midst of the modern world.) Yet, when it came to social problems, Wells became a votary of scientism.

One of the most poignant examples of such ambivalence was Bertrand Russell, amongst the first to sense the full destructive power of modern science and technology. In his *Icarus*, an essay on the future of science, as well as in a number of other works, Russell touched upon the relationships between authoritarian control, science and technology, and the instrumental use of isolated rationality. As a corrective, he wanted both

[16] T. Kroeber, 'The Coping Function of Ego Mechanisms', in R. W. White (ed.), *The Study of Lives* (New York: Atherton, 1963), pp. 178–98.

reason and love, not isolated reason.[17] Yet, in his system, reason had an intrinsic legitimacy; love did not. Love had to be reasoned love; reason did not have to be feeling reason. He wanted love *and* reason, not love *in* reason.

At least two millennia before modern psychology was born, the *Kaushitaki Upanishad* advised one to try to understand the speaker behind the spoken word and the doer behind the deed.[18] And I hazard the crude *ad hominem* argument that Russell's own life provides clues to the disjunction between ideas and feelings that his philosophy endorsed. His emotional relationships showed that he never sensed the subtle exploitation in a two-person situation where one operated according to the principle of rational love and the other had faith in reasons of heart. He never imagined that what Freud might have called a rational transference could become—with its built-in bias for impersonal, negotiable, part-object relationships—an instrument of oppression. The simple, non-intellectual biography of Russell by his daughter Katherine Tait recognizes this. It unwittingly reveals how Russell's own children rebelled against the oppression of rational love. Katherine herself found religion and missionarism, both as a means of de-isolating and as a means of defying her aggressively atheistic father; and her brother found madness, of a kind which usually has the split between ideas and feelings as its main symptom.[19] It is Mrs Tait's naïve comment in the context of her brother's illness which turns out to be *intellectually* the most challenging; she in effect wishes that her father had been more influenced by the open-ended, easily criticizable, more holistic and less scientific psychology of Sigmund Freud than by the positivist, progressive and ultra-scientific system of J. B. Watson.

[17] See the 'Prologue' in Bertrand Russell, *Autobiography* (London: Unwin Paperback, 1975), p. 10.

[18] 'Kaushitaki Upanishad', translated with comments by Prafullakanta Basu, in *Upanishad*, Vol. 2, ed. Sitanath Tattwabhushan, trans. and commentary by Maheshchandra Vedanta-Ratna, and Prafullakanta Basu (Calcutta: Haraf, 1976), 2nd ed., pp. 511–77; see pp. 563–64.

[19] Katherine Tait, *My Father Bertrand Russell* (New York: Harcourt, Brace, 1975), pp. 62–4.

Implicit in such torn creative minds of this century's Europe was the belief that while the context of modern science and its applications were faulty, the text of science was liberating. In fact, as diagnosed by the modernists, the problem was that the objectivity of science had not yet fully informed the social uses of science. That is, while the scientifically minded had used isolation, they had not isolated deeply and widely enough; feelings still dominated many sectors of human life, and these sectors were waiting to be liberated by the further growth of the scientific temper.

Some years ago Gerald Holton, one of those optimists who are not embarrassed to seek security by surrendering more fully to the forces which cause the insecurity in the first place, declaimed:

While we may intuitively feel that the choice is unpleasant, it is perhaps not necessarily so paradoxical as it seems. A number of social or physical systems offer models in which stability, when disrupted by the introduction of a new factor, can be reestablished at some level only by increasing the role of the new factor even further.[20]

Predictably, a majority of natural scientists toed this line. Not so predictably, many social analysts, too, chipped in with the same analysis. They valiantly tried to solve the social problems of science by promoting more science. The new credo was: the content or text of modern science is universal and amoral but its social context is often parochial, value-loaded and evil. Individual scientists, too, can sometimes be self-interested, hypocritical or opinionated. Change the social relations of science and you will have finally an ethically pristine, fully liberating, modern science.

Entire schools of thought have by now grown up on this staple diet, and the Ernest Nagels and Peter Medawars have even tried to build an entire dietetics on it. As such ideas and their political power are widely known, I shall not discuss them further. Instead, I shall draw attention to the new generation of ordinary citizens and consumers of science who have

[20] 'Introduction', in Gerald Holton (ed.), *Science and Culture: A Study of Cohesive and Disjunctive Forces* (Boston: Beacon, 1965), p. x.

been so well brought up on the principle of the purity of scientific texts that they, even when practicising homeopathy or palmistry or even when growing a sacred tuft of hair or going on a pilgrimage, have to justify themselves on scientific grounds. Among the third-world elites today, such uncritical acceptance of science as the absolute standard of validation is now more common than the Asian flu.

This growing body of uncritical supporters of science operate with the same folk philosophy with which, according to Bruno Bettelheim, apolitical victims often face oppression in 'extreme situations'. Used to being obedient to the scientific establishment, they dare not oppose the ruling ideology. Each inhumanity imposed or legitimized by science is seen as a mistake of the system which could be corrected from within it. [21]

IV

Today, in the last quarter of the twentieth century, another response is conceivable. Older, tired and wiser, we can now take courage to affirm that the main civilizational problem is not with irrational, self-contradicting superstitions but with the ways of thinking associated with the modern concept of rationality; that modern science has already built a structure of near-total isolation where human beings themselves—including all their suffering and moral experience—have been objectified as things and processes, to be vivisected, manipulated or corrected. According to this view, the irrationality of rationality—as Herbert Marcuse might have described the pathology—in organized normal science—as Thomas Kuhn might have described the system—is no longer a mere slogan. It is threatening to take over all of human life, including every interstice of culture and every form of individuality. We now have scientific training in modern sports and recreations; our everyday social relations and social activism are more and more guided by pseudo-sciences like management and social work and by pseudo-technologies like transactional analysis and T groups.

[21] Bettelheim, *Surviving.*

Our future is being conceptualized and shaped by the modern witchcraft called the science of economics. If we do not love such a future, scientific child-rearing and scientific pedagogy are waiting to cure us of such false values, and the various schools of scientific psychotherapy are ever-ready to certify us as dangerous neurotics. Another set of modern witch-doctors has taken over the responsibility of making even the revolutionaries among us scientific. In fact, the scientific study of poverty has become more important than poverty itself. Even in bed, our performance is now judged according to the objective criteria of some highly scientific, how-to-do-it manuals on sex.

Such a process has continuously justified our ability to freeze or fix a subject for study and to place it at a distance to evaluate. Those acquainted with Bettelheim's account of human beings facing arbitrary torture and murder will know why I have used the word 'distance' here. Distancing is a psychological device which both the victim and his oppressor have to use, one to ward off the reality of his fate and the other to reduce his victim into an object.[22]

It is the second use which is pertinent to my argument here. It is the use which prompts Aimé Césaire to write the quaint formula: 'colonization = thingification'.[23] In its extreme form such objectification becomes necrophilia, the passion to kill so as to freeze, place at a distance, and love.[24]

The warning against the rationality from which the objectification derives is best given in the words of Fromm:

Logical thought is not rational if it is merely logical. . . . (Paranoid thinking is characterized by the fact that it can be completely logical . . . Logic does not exclude madness.) On the other hand, not only thinking but also emotions can be rational. . . .

Reason flows from the blending of rational thought and feeling. If the two functions are torn apart, thinking deteriorates into schizoid

[22] Ibid., Part 1.

[23] Aimé Césaire, *Discourse on Colonialism*, trans. Joan Pinkham (New York: Monthly Review Press, 1972), p. 21.

[24] Erich Fromm, *Anatomy of Human Destructiveness* (Connecticut: Fawcett, 1973). See also George Devereux, *From Anxiety to Method in the Behavioural Sciences* (The Hague: Mouton, 1967).

intellectual activity, and feeling deteriorates into neurotic life-damaging passions.

The split between thought and affect leads to a sickness, to a low-grade chronic schizophrenia, from which the new man of the technetronic age begins to suffer. . . . There are low-grade chronic forms of psychoses which can be shared by millions of people.[25]

Fromm here endorses, with the help of nosological entities similar to the ones I have used, the social analyses which nervously view a growing number of societies getting structurally and morally reorganized to meet the needs of organized science. He in the process unwittingly provides another reason why criticisms of modern science from within the scientific worldview cannot go very far.

The importance of the other position which insists that the social problems created by modern science cannot be handled within the culture of modern science, has also grown because the idea of more science to cure the ills of science seems especially to enthuse normal scientists and the political spokesmen of the scientific estate. It is now obvious that the slogan of internal criticism and the search for the hair of the dog to cure dog-bite serve the interests of the scientists rather well, for they delegitimize criticisms from the outside and suggest that while the scientific worldview cannot be judged by other worldviews, the other worldviews can be judged and indeed should be judged by science.[26]

To give a well-known example, Paul Feyerabend, no lover of astrology himself, examines at one place a joint statement by 186 modern scientists, eighteen of them Nobel-laureates, against astrology.[27] He shows that none of the 186 had studied astrology before attacking it. Some of them, when contacted by journalists, were unashamed that they knew nothing about astrology. Their statement shows the same ignorance of the relevant find-

---

[25] Erich Fromm, *The Revolution of Hope: Toward A Humanized Technology* (New York: Harper and Row, 1974), pp. 42–3.

[26] I have discussed this issue in more detail in 'Science in Utopia: Equity, Plurality and Openness', *India International Centre Quarterly*, 1983, 10(1), pp. 47–59.

[27] Paul Feyerabend, 'The Strange Case of Astrology', *Science in a Free Society* (London: NLB, 1978), pp. 91–6.

ings of modern science. That of course did not stop them from passing judgement. Not only were they unwilling to apply their scientific method to judge the claims of a competing system, they did not stop to ask why they needed 186 signatures and not one, if the arguments were so good and so conclusive.

One is tempted to argue that the 186 signatures were necessary mainly to deny the principle of reciprocity. They were meant to deny the counter-claim that, if modern science claims the right to criticize other systems, it should give the right to criticize science if not to other systems at least to its own victims, that it should grant that a part of the ethical restraint on modern science may now have to come from outside science, from the totality of human experience confronting science.

Any idea of external control on science, however, sounds like a denial of free thought to many. Discredited by the clumsy, sometimes tragic battle waged against science by the medieval church, the idea of external control seems dangerous even now, when science rules the world. But could it be that the church in its obscurantism was expressing its fears of a system of knowledge freed from the restraints of ethics and social conscience, however faulty that ethics and however rigid that conscience? The answer may be less unfriendly to the church today when modern science is a part of the global establishment, when most faiths have become defensive and all organized faiths are seeking endorsement from science. Today the issue is: which pathology has become more unsafe for human survival, that of scientific rationality or that of its 'irrational' subjects?

V

The problem I am posing is, I hope, clearer. I am suggesting that when the world of uncritical traditions faced the first onslaught of organized modernity, the principle and practice of isolation played a major role in it. Modern science at that stage was a creative, and modern authoritarianism a pathological, possibility of the ability to isolate. Gradually, over-isolating, fully organized modern science has become another pathological correlate of the demise of traditions and the erosion of cultures, the false claims of the rationalists, scientific socialists and Hobbesian liberals notwithstanding.

The earlier creativity of modern science, which came from the role of science as a mode of dissent and a means of demystification, was actually a negative force. It paradoxically depended upon the philosophical pull and the political power of traditions. Once this power collapsed due to the onslaught of modern science itself, modern science was bound to become, first, a rebel without a cause and then, gradually, a new orthodoxy. No authority can be more dangerous than the one which was once a rebel and does not know that it is no longer so.

The moral that emerges is that modern science can no longer be an ally *against* authoritarianism. Today it has an in-built tendency to be an ally *of* authoritarianism. We must now look elsewhere in the society to find support for democratic values.

Why has something which began as a movement of protest become part of the Establishment? Why do the moderns continue to view science as a cornered voice of dissent fighting powerful opponents when it all too visibly owns the world? Why do even the radical critics of society exercise restraint when criticizing science?

Any answer to these questions must begin with the admission that modern science is both a social institution and a search for new meanings and aesthetics. During its first two centuries, it was the second aspect of modern science which predominated. In Europe till the eighteenth century the scientist was claiming the right to search for another truth and adopt another mode of reaching it. But that philosophical quest was a hangover from the days of classical science and the scientists recovered from it soon enough to produce, by the end of the nineteenth century, a formidable organization and strong links with that other child of seventeenth-century Europe, the modern nation-state system. In another five decades, the scientist has become the main author of the Establishment cosmology. He is now the orthodoxy; he is now the Establishment. So much so that to perceive him still as a weak, unorganized fighter against authority can spell disaster for all of us.

When science was primarily a philosophical venture, it allowed for more plurality. In the days of organized science there is little scope for a scientist to protect his individuality as a scientist. Overorganized science has managed to do the impossible: it has become a market-place and a vested interest at the same time. It has an organizational logic independent of the creativity of the individual scientist but dependent on—and subserving—his material interests. Thus, there is an inner incentive for the scientists—for even the most creative among them—to orient their creativity to the dominant culture of science. The scientist can fully encash his creativity in the market-place of science only if he plays according to the existing organizational rules of modern science and, better still, if he remains unconscious of the rules in the fashion of what Georg Lukacs calls the silent species.[28]

This depoliticization is camouflaged by a special brand of pseudo-politics. The normal scientist, who could be defined as the practitioner of Thomas Kuhn's normal science, is expected to be politically involved, but he is expected to operate *as if* the pathology of modern science lay only in its context. He can shout himself hoarse over nuclear armaments—as a pacifist, a liberal or as a Marxist—but he cannot say that violence lies at the heart of modern science. He may speak of the origin of science in superstitions, prejudices and myths; he can speak of the persistence of these in the individual scientist; but he cannot speak of their persistence in the text of science. In other words, there is now a standard officially-sponsored model of political dissent for the scientists. If a normal scientist follows that model, science rewards him handsomely, otherwise he is valued not as an eccentric professor but as a lunatic who has missed his professional bus. It is this cultural twist which has pre-empted basic internal criticism in science.

This point can be made in another way. The culture of

[28] Georg Lukacs, 'The Twin Crisis', in San Juan, Jr. (ed.), *Marxism and Human Liberation* (New York: Delta, 1973), p. 316.

modern science gives a special role to the scientist in defining the concerns of science, whether these concerns be textual or contextual. But it encourages him to shirk all responsibility if something goes wrong with the concerns. That responsibility is passed on to other citizens. Thus, the scientist gets the credit for the constructive discoveries of science, not for the destructive ones. Indeed, his training encourages him to either criticize science only in terms of its context ('Nothing is wrong with nuclear research; the politicians and the generals are the ones who misuse it and produce nuclear arms') or reduce all contextual problems to textual ones ('If science threatens an ecological disaster, do not seek woolly social or political solutions; seek scientific ones, for science can always solve the problems it has created').

This is the other way the culture of science is structured by ego-defence like isolation and denial, and controlled by a small number of two-dimensional scientists who, unlike the political elites, have exempted themselves from criticisms, checks and competition. The bureaucratic violence that results is endorsed by the total socialization of the individual through modern child-rearing, education and mass media. The scientists decide the use of science in society; the lay person considers such control proper. Increasingly, scientists exercise their power with the enthusiastic approval, in fact on the demand of a section of the society. Both sides view the suffering inflicted by or in the name of science as a needed sacrifice for the advancement of human rationality and social progress.

The traditional cultures, not being driven by the principles of absolute internal consistency and parsimony, did allow the individual to create a place for himself in a plural structure of authority. In such cultures the individual always had some play vis-à-vis the institutions he worked with. For instance, a guru could be a false consciousness to many but, traditionally, one man's guru was always another man's anti-guru. Such fragmentation of the world of gurus was presumed by every disciple of every guru. So there were at least varieties of false consciousness competing for the allegiance of the believers.

Such multiplicity is not granted by modern science which, because it presumes universal norms and unitary truths, must reject all gurus, and claim religious allegiance to one truth and one form of liberation. So you have faith but faith without the different forms of godmen, revelations and prophets which enriched the traditional religions.

Finally, the four pluralities science disarmingly accepts. In each case, there is an implicit but irrevocable principle of hierarchy as well as a totalist vision of social consciousness. First, there is classical science, by which one means pre-modern Western science, seen as a heroic, but an earlier, romantic and inferior stage in the evolution of true knowledge, the final stage of which is presumed to be modern Western science. In this hierarchy classical science is fitted in as a museum-piece, not as an alternative view of nature and humanness.

Second, there are the ethnosciences, the non-modern, non-Western traditions of science which are seen as semi-scientific reservoirs from which modern science may have to pick up insights and practices, rejecting the rest as so much mythology and magic. The borrowing by modern medicine of reserpine from Ayurveda does not imply any respect for the philosophy or the structure of Ayurveda; it shows a pragmatic openness towards some specific findings of Ayurveda. It is the respect we show an alert child who by chance spots a misplaced railway ticket which the elders should have spotted in the first place but, through a series of accidents and oversights, did not.

Third, there is the internal plurality of competing scientific theories. It, too, has no intrinsic legitimacy. If science has more than one explanation of a phenomenon, the expectation is that only one of them will finally win and establish its hegemony; otherwise a new theory will emerge and supplant all the competing theories. Usually, of course, there is one dominant theory in existence; this is held by the scientists in the fashion of, to use Kuhn again, a totalizing dogma.

The fourth plurality, too, is internal. Scientists grant legitimacy to the divide between what J. R. Ravetz calls the mature and the immature sciences.[29] Though theoretically any kind of science can be immature, in practice the social sciences are so classified, mainly because of their paradigm-surplus nature. For all paradigm-scarce disciplines are definitionally mature following Kuhn. This is despite the critical power the human sciences sometimes derive from their paradigm-surplus nature and from their ability to offer wider social choices as well as openness of vision.[30] The main function of this concept of maturity is to avoid having critical social sensitivity close to the heart of science.

The pluralities of science, therefore, are no pluralities at all. They may be necessary for the progress of modern science but to participate in or manage such a culture of science requires something more than the qualities imputed to the stereotypical scientist; they require a complex of psychological skills most frequently found in the authoritarian personality, either as part of a search for 'authoritarian domination' or as an expression of 'authoritarian submission'.

VI

I have said that modern science was once a movement of dissent. It then pluralized the world of ideas. I have said that it is now the centre-piece of the Establishment cosmology and can function neither as an instrument of basic criticism nor as an expression of scepticism—its philosophical hallmarks at one time. I have also said that modern science, at its best, was once a creative response to a particular psychological problem, the pathological response to which later turned out to be modern authoritarianism. I am now suggesting that modern science, which began as a creative adjunct to the post-medieval world and as an alternative to modern authoritarianism, has itself acquired many of the psychological features of the latter. In

[29] J. R. Ravetz, *Scientific Knowledge and Its Social Problems* (Harmondsworth: Penguin, 1973), pp. 156–9.
[30] The problem of pluralities has also been discussed in Ashis Nandy, 'Science in Utopia', *India International Centre Quarterly*, 10 (1), pp. 47–59.

fact, in its ability to legitimize a vivisectional posture towards all living beings and non-living nature, modern science is now moving towards acquiring the absolute narcissism of a new, passionless Caligula.

Modern science began by giving a dissenting meaning to the man–nature relation. It was not merely another ideology claiming that other ideologies were false or inferior; it was another view of the human condition which sought to make all ideologies redundant. (The end-of-ideology argument, so popular a decade ago, can be seen as a projection of the triumph of this anti-ideology in human mind and society.) In its earliest form, modern science disturbed the older world image not so much by being unconditionally true, but by introducing a new style of demystification which subverted parts of the European tradition that had become stale, self-justifying and inconsistent with experience. This is why when specific scientific theories were falsified and reduced to the status of myths by the growth of modern science, it did not lead to any great jubilation among the believers, not even when the falsified theories dealt with matters of theological concern. The believers sensed that modern science had offered a way of looking at things which was partly independent of the changing content of modern science. They sensed that one could not escape the critical gaze of modern science by taking advantage of the changes within it.

However, like some of the schools of social criticism it directly or indirectly spawned, modern science too developed features which were to help it, as a critical tradition, to demand and get uncritical support. Not only did modern science gradually develop a rigid, unidirectional mode of demystification which saw all such other modes as subsidiary or peripheral, it began to see all alternatives to its mode of demystification as conspiracies against human good. This was backed up by a self-justifying tough-mindedness.[31] What was first a quality of consciousness was now institutionalized and concretized as a 'thing' and as

[31] The word has been borrowed from modern psychology which uses it to distinguish indirectly the more scientific from the less, and the better from the worse.

an independent reality, in fact the only reality.

First, there was the concretization of concepts. Rationality, for instance, was once an attribute of thinking. It became a concrete body of knowledge and a set of methods of knowing. Adjectives thus became nouns and the psychological became the crypto-physical under the influence of an anti-intraceptiveness which, in another context, was later found to be closely associated with modern authoritarianism.

Second, the worlds of nature and, later on, human nature came gradually to mean the worlds of the sciences of nature and of human nature. This is not the old argument about science cornering culture, though that argument, too, has some power. I am speaking of the operationalism which reduces reality to the reality accessible to the methods of science, and then reconstructs the 'whole' reality—of nature, persons or cultures—by extrapolating from that operational reality. The dangers of such concretization—and the isolating, part-object relations it promotes—are especially obvious in the human sciences. In psychology, for instance, intelligence tests are no longer seen as imperfectly operationalizing intelligence; intelligence is now what the intelligence tests measure. A strategy of research has come to define the whole of the reality of human intellect.

I am often told that this is a price we must pay for the growth of science, and once the infant science of psychology matures, it should be able to handle the complexities of human nature. I am not so sure. The rewards of operationalism and that of the control it gives over individuals and groups are enormous. And once it is institutionalized in a society, it acquires more and more autonomy. The means gradually begin to define the goals and ultimately become the goals. In another context, Freud might have called this an instance of process pleasure—the pleasure which should be associated with an instinctual goal but is displaced on to the process of reaching the goal.

Finally, within the scientific estate there is the pressure on objectivity to move closer to objectification due to the constant stress on the subject–object dichotomy. In the modern knowledge systems, this dichotomy is seen as a major pathway to power through knowledge and to knowledge through isolation.[32] This has necessarily led to a further endorsement of mechanomorphism. The old European concept of the world machine included the idea of God the clock-maker which, retrogressive though it may sound to modern ears, did provide a check on the potential for isolated cognition implied in the idea of the world as a machine. The new secularized concept of the world machine represents a desacralized mechanomorphism which admits no limit on itself. Behaviourists like J. B. Watson and B. F. Skinner have only taken to its logical conclusion this process of objectification. How far they derive their legitimacy from the promise of scientific control over human fate is obvious from the fact that behaviourism remains the official ideology of both the orthodox modernism of the West and the critical modernism of Soviet Marxism.

Any mention of the duality of the observer and the observed prompts a section of scientists and philosophers of science to mention particle physics, Werner Heisenberg or microbiology. And then some social scientists join them with Freud's concepts of transference and counter-transference or the structuralist concept of the savage mind. As if these concepts defined the mainstream culture of modern science or disturbed the poise of the normal scientist pursuing his normal science! I do not think it an overstatement to say that the culture of normal science, as we know it, will collapse if it gives up the division between the observer and the observed or the hierarchy between the scientists and the laity.

Once again we are close to what some psychologists have identified as a basic feature of political authoritarianism: all-round objectification and the idea of a leadership supposedly representing both the true interests of the masses and the superior understanding of those interests. Political authoritarianism *has* to see the citizen as a subject whose subjecthood is

---

[32] Gregory Bateson is one of the many who have suggested that the objectivity of experience is a typically Occidental view of the world. See his *Mind and Nature: A Necessary Unity* (Toronto: Bantam, 1980), pp. 33–4.

no different from that imposed on the laity by science. The sometimes harmless distance between the scientist and his subject becomes in politics the chasm between a self-declared elite —the 'revolutionary vanguard' in some theories of progress— and their increasingly voiceless objects of manipulation: the reportedly immature masses, underdeveloped, primitive, and carrying the heavy baggage of false consciousness. Seen thus, the culture of modern science is part of a more general theory of imposed secular salvation, the other special case of which is modern authoritarianism.

It is therefore not a paradox of our times that to contain modern science many are falling back on what has been one of the main targets of modern science during the last three hundred years—cultural traditions. It is part of the attempt to protect the plurality of human consciousness and provide a critique of science from outside. In so far as the various in-house criticisms of modern science have not defied modernity and in so far as modern science is inextricable from the modern consciousness, in many societies one is forced to fall back on the traditional worldviews. At least the latter have tried to protect, at the margins of the 'civilized' world, the crucial insight that the battle against isolation is joined when one gives up the concept of a fully autonomous observable and opts for the dyad of the observer and the observed as the basic unit of analysis. A number of non-modern systems of thought have sought freedom and understanding in the deliberate search for a continuity between the observer and the observed, in cross-identifications and empathy. Here, for example, is Toshihiko Izutsu speaking of Islam:

The problem of the unique form of subject–object relationship is discussed in Islam as the problem of *ittihad al-alim wa-al-ma' lum*, i.e. the 'unification of the knower and the known'. Whatever may happen to be the object of knowledge, the highest degree of knowledge is always achieved when the knower, the human subject, becomes completely unified and identified with the object, so much so that there remains no differentiation between the two. For differentiation or distinction means distance, and distance in cognitive relationship means ignorance.[33]

True, the traditional philosophies generally place such unity of the knower and the known outside everyday life, which these philosophies often see as unavoidably dualistic. Nonetheless, the awareness of such possibilities delimits the role of modern science and helps one to see it as only a finite system of knowledge and as a corrective to an overly projective worldview. Such delimitation in turn allows the peripheries of the world to use their traditions as a legitimate vantage ground for social criticism.

This, however, only brings us to another question: what kinds of tradition can be used as tools of criticism and what kinds are open to criticism? Apparently, the answer to this question is known. One knows the kind of tradition which renaissance science criticized and the reasons thereof. The moderns never tire of remembering the isolating, heartless, frozen aspects of traditions which Galilean science attacked. Modern Indians, too, never fail to remind themselves that the last two hundred years of Indian life have been a continuous struggle against not merely the colonizing West but also the negative aspects of Indian traditions. Even the counter-modernists grant that cultural traditions can become ritualized, self-justificatory and a means of perpetuating institutionalized violence. They grant that traditions, too, may push one to isolate their contents. It is probably in the nature of any complex cultural system to seek self-perpetuation through isolation. After all, according to Freud, the main role of rituals is to isolate, and a culture is hardly conceivable without its own quota of rituals.[34]

This is only another way of saying that no culture can survive on a staple diet of passions. Nothing can be as dead as last year's passions. A culture must constantly persevere, if that is the word, to survive on an appropriate mix of non-heroic

[33] Toshihiko Izutsu, *The Concept and Reality of Existence* (Tokyo: The Keio Institute of Culture and Linguistic Studies, 1971), p. 5.
[34] Freud, *Inhibitions, Symptoms and Anxiety.*

self-definition and ritualization of everyday life.

Let us not, however, minimize the complexity of the problem. Choosing the right traditions is not a matter of choosing from among the discrete elements of a culture. A culture is not a grocery store, with each customer a free purchaser and each purchase an independent purchase. A culture is an interconnected whole with some strong interconnections and some weak; a culture has some odd, unpredictable, ill-understood bonds with those who live by it, use it or even disown it. Within it, you have some options only if you exercise others, and the options exist only if yet others are not exercised. The choice of traditions I am speaking of involves the identification, within a tradition, of the capacity for self-renewal through heterodoxy, plurality and dissent. It involves the capacity in a culture to be open-ended, self-analytic and self-aware without being overly self-conscious.[35] There are traditions, or at least constructions of traditions which, even when you introduce crucial changes into them, are not threatened. These traditions can give meanings to the changes in terms of categories internal to them. Because they have subtraditions which operate as baselines for social criticism, they are accustomed to converting external criticisms into internal ones. On the other hand, there are traditions which are so fragile or so consistent internally that the removal of a single plank may mean total collapse. In neither case can one mechanically apply the principle of choice.

Fortunately, cultures are usually more open and self-critical than their interpreters. In the first half of this century, Ananda

[35] Apart from my obvious indebtedness to the critical tradition, I have in mind here the meaning of 'analysis' that emerges from the works of Philip Rieff on Freudian ethics. See especially his *The Triumph of the Therapeutic: The Uses of Faith After Freud* (New York: Harper, 1968). Such a meaning in some ways ties up with the concept of criticism as used throughout this paper. Though neo-Freudian and neo-Marxian in origin, the concept does have some degree of cross-cultural validity. It certainly ties up with the critical uses to which some forms of *advaita*, especially the theory of *maya*, could be put. Also relevant in this context is the work of one who may seem a strange bed-fellow, Karl Popper. See his 'Towards a Rational Theory of Tradition', in *Conjectures and Refutations: The Growth of Scientific Knowledge* (London: Routledge and Kegan Paul, 1972), pp. 120–35.

Kentish Coomaraswamy wrote his brilliant critique of the modern civilization. He contrasted this civilization with the traditional vision of man—humane, contemplative and just. He thus took to an elegant conclusion the critique initiated by Thomas Carlyle, John Ruskin, William Blake and Leo Tolstoy on the one hand and a galaxy of non-Western thinkers on the other. However, even if one grants that everyone has the right to project a utopia into the past, Coomaraswamy's tradition remains homogeneous and undifferentiated from the point of view of man-made suffering. His defence of the charming theory of *sati*, for example, never takes into account its victims, the women who often died without the benefit of the theory. By refusing to consider this mundane issue, Coomaraswamy's traditionalism ceases to be critical, however open it might be metaphysically to the idea of self-criticism and self-renewal. Such traditionalism reactively demystifies modernity to remystify traditions.[36] It also promotes isolation, even if in a much less dangerous form than did Dr Josef Mengele and Shiro Ishii under the banner of science.

Likewise, one may concur with Coomaraswamy that the untouchables in traditional India were better off than the proletariat in the industrial societies. But this could be an empty statement to those victimized by the caste system today. When many untouchables opt for proletarianization in contemporary India, is their choice merely a function of faulty self-knowledge? Can we draw a clear line between the experts on traditions and the laity, and declare the latter's knowledge, feelings and values irrelevant to the understanding of traditions? Are we not then replicating nineteenth-century colonial anthropologists and historians who stratified persons, races and cultures into the producers and the consumers of knowledge, into those who were historians to the world and those who were objects of history? I am afraid Coomaraswamy's traditionalism, despite being holistic by design, does not allow a creative, critical use of modernity within traditions. This never happens with the living traditions which Coomaraswamy theoretically supports. The

[36] See 'Evaluating Utopias' in this volume.

Ramayana and the Mahabharata, for instance, take into account the modern consciousness in the form of the personality types represented by some demons (*danavas, daityas, rakshasas* and *asuras*), and some anti-heroes such as Karna. These types are rejected; but they are first considered seriously, given due respect and used as critiques of the types favoured.

An excellent example of the critical use of modernity within tradition is the two hundred years of the recent past of Indian society from Rammohun Roy to Gandhi. Throughout the period, continuous and sometimes successful efforts were made to make the modern world a meaningful—and manageable— part of Indian experience. Even the parallel negative past of modern India—from Radhakanto Deb, who opposed Rammohun Roy, to Nathuram Godse, who killed Gandhi—can be read as an unsuccessful effort to arrive at a creative use of modernity. That such efforts did not always succeed or that they often led to dangerous visions should not blind us to the seriousness of the efforts. Deb opposed the abolition of *sati* by the British, but was a pioneer in women's education. Godse was an ultra-Hindu, but the Hinduism he fought for was more modern than Gandhi's. A part of Coomaraswamy's problem arises from his emphasis on the classical at the expense of the folk and on the 'pure' at the expense of the 'hybrid' and the 'dirty'. Perhaps if he had not had that odd middle name, if he had not had to disown his mixed origin and bicultural consciousness, or live away from his tradition for so long with such enormous knowledge of it, he might have defended Indian culture less uncritically.

Today, with the renewed interest in cultural visions, one has to be aware that commitment to traditions, too, can objectify by drawing a line between a culture and those who live by that culture, by setting up some as the true interpreters of a culture and the others as falsifiers, and by trying to defend the core of a culture from its periphery. Such uncritical commitment tends to undervalue the folk as opposed to the classical, the contextual as opposed to the textual, the reinterpreted as opposed to the professionally interpreted, and the subsequent or 'interpolated' as opposed to the earlier or the 'original'. As in science, so in culture. A closed system tends to become a vested interest, sometimes in the name of openness.

Some of the models of Hinduism produced during the last one hundred and fifty years neatly exemplify the consequences of such onesidedness. They glorify Hinduism but tend to look down upon the Hindus. Thus, Swami Vivekananda's tradition-alism defended the texts and symbols of Hinduism fully but sought to improve the Hindus by giving Hinduism an institutional structure borrowed from Western Christianity. Though he attacked some of the Westernized reformers of Hinduism, he also sought to create, by his own admission, a Western society of Vedantic Hindus to pay back the Imperial West in its own coin.[37]

Vivekananda, like Bankimchandra Chatterji before him and Bal Gangadhar Tilak after, sought to blend with Hinduism elements of positivism, socialism, nationalism and masculine Christianity, including the Protestant work ethic. This spirit of synthesis has played, for better or for worse, a significant role in Indian politics for nearly one hundred years.[38] The other versions of Hindu nationalism have been cruder; they have devalued the living Hindu and sought to improve his character and potency, to turn him into a proper counterplayer—often a mirror image—of the conquering Westerner and the 'potency-driven' Muslim. In its self-hatred, Hindu nationalism has wanted to rewrite Hinduism as a 'proper' religion, as well-organized and well-bounded as organized Christianity and Is-lam. The ordinary Hindu probably senses the threat to his survival posed by such cultural engineering; politically, Hindu nationalism had been reduced to an urban, semi-modern, middle-class phenomenon.[39]

[37] Ashis Nandy, *The Intimate Enemy: Loss and Recovery of Self under Colonialism* (New Delhi: Oxford University Press, 1983), chapter 1.

[38] Ibid.; also 'The Making and Unmaking of Political Cultures in India', in my *At the Edge of Psychology: Essays in Politics and Culture* (New Delhi: Oxford University Press, 1980), pp. 47–69.

[39] A pathetic expression of this ideology was Nathuram V. Godse, the assassin of M. K. Gandhi. For an analysis of the clash between two forms of Hinduism

The psychogenesis of such nationalism has been explored in depth in Rabindranath Tagore's novel *Gora*, which tells the story of an ultra-Hindu who turns out, at the end of the novel, to be the abandoned child of an English couple. An accident of life history here symbolizes a deeper cultural equation: the more doubtful one's roots, the more desperate one's search for security in exclusion and in boundaries. Gora, however, proves himself more authentic than those he symbolizes. At the end of the novel he opts for the wisdom of a more inclusive consciousness, not as a compromise but as a superior form of Hinduism.

Tagore here is hinting at another kind of tradition which is reflective as well as self-critical, which does not reject or bypass the experience of modernity but encapsules and digests it. Such a tradition refuses to give primacy to the needs of pure cognition at the expense of totality of consciousness and it refuses to sanction total redefinition of itself in response to defeat or humiliation. It of course rejects imitation, but it goes beyond that and rejects, as a path to self-esteem, the compulsive rejection of other cultures and fights the compulsion to be only the other culture. Even in defeat, it retains its authenticity, though it incorporates the experience of defeat as relevant.

Not being a Gandhian, I can say without any apologia that Gandhi represented such a concept of critical traditionalism aggressively. (Tagore recognized this, and though he had reservations about many aspects of Gandhism, it was the Gandhian theory of nationalism which he found least offensive.) Not being a Maoist, I can afford to say, now that the semi-educated peasant is no longer in fashion, that in some of his incarnations he probably had an inkling of what was involved in such rootedness. He attacked Confucianism, but, often against himself, he sought to fit Marxism within Chinese culture rather than the other way round.

Not being a Marxist, I shall only hesitantly say that Marx himself was often a prisoner of nineteenth-century scientism

_____

protesting differently against colonialism, see Ashis Nandy, 'The Final Encounter: The Politics of the Assassination of Gandhi', *At the Edge of Psychology*, pp. 70–98; and 'Godse Killed Gandhi?', *Resurgence*, January–February 1983, (96), pp. 28–9.

and the petty ethnocentrism it underwrote. In spite of his seminal contribution to the demystification of the industrial society, he had no clue to the role modern science had played in legitimizing such a society and in the repression of other cultures and societies.[40] (And if one is not sensitive to the way science has provided a model of domination in our times, one cannot be sensitive to the way the non-modern cultures can provide a baseline for social criticism.) A faithful product of Enlightenment, Marx acquitted science and put it outside history, locating the source of human exploitativeness solely in the sphere of political economy. It is thus that his theory kept the door open for scientific social engineering based on objectification of persons and groups. That is why Stalin is not an accidental entry in the history of Marxism. He remains a brainchild of Marx, even if, when considered in the context of Marx's overall vision, an illegitimate one.[41]

The critical traditionalism I am talking about does not have to see modern science as alien to it, even though it may see it as alienating. It sees modern science as part of a new cognitive order which can be occasionally used for critical purposes within the earlier traditions. Such traditionalism uncompromisingly criticizes isolation and the over-concern with objectivity, but it never denies the creative possibilities of limited objectivity.

Wisdom recognizes continuities as much as change; it recognizes optimality and the limits of applicability of concepts and character-traits. As in the clinic, so in the culture. Ultimately, intelligence and knowledge are poor—in fact, dangerous—substitutes for intellect and wisdom.

VII

I might be able to make my point better by recalling a brief, apparently trivial, episode in the life of M. N. Roy. It is said

[40] A third-generation Marxist like Jürgen Habermas has done better in this respect. See his 'Science and Technology as Ideology', in *Toward a Rational Society* (London: Heinemann, 1977), pp. 81–122.

[41] See on this subject Leszek Kolakowski, 'Marxist Roots of Stalinism', and Mihailo Marcovic, 'Stalinism and Marxism', in Robert C. Tucker (ed.), *Stalinism:*

that once when he was ill during his last days, Roy insisted that his wife Ellen wear, while nursing him, a red-bordered white sari as his mother used to do in his childhood. Others have disputed the veracity of the story. Being rationalists, they evidently see the irrationality of any rationalist as dangerous spicy gossip. That a person may not choose to work with objectivity in all situations seems to them not merely vulgar; it is a fall from humanness itself.

But should objectivity work in all cases? I like to believe that when Roy reportedly 'fell' from his rationalism by seeking a symbolic reaffirmation of his private concept of motherhood and mothering, he was actually admitting the continuities in the symbols of nature and *caritas*. Perhaps against his will, he admitted some of the undying concerns of his culture and the subtler modes of cultural communication among human beings who are ready to 'listen'. That is, he accepted the limitations of the conventional concept of rationality and tried to be true to the full meaning of his own faith—that human reason and morality expressed the harmony of the cosmos.[42] That is why Roy wanted from his wife not only professional nursing and the institution called medical after-care, but wanted these hard realities to be given meaning with the help of the traditional symbols, and the feelings and aesthetics associated with them. He was recognizing the mysteries called maternity and wifeliness, and accepting Thomas Mann's maxim that 'It is love, not reason, which is stronger than death.' He was de-isolating.

I want to believe that this disputed episode in Roy's life is true. To admit such an episode is to admit that Roy was, through his apparent irrationality, expressing his superior intellect and his superior wisdom, if not a higher form of rationality itself.

Essays in Historical Interpretation (New York: Norton, 1977), pp. 283-319. On the roots of technocratic Marxism in the positivist Marx, see Albrecht Wellmer, *Critical Theory of Society* (New York: Herder and Herder, 1971).

[42] M. N. Roy, *Reason, Romanticism and Revolution*, Vol. 2 (Calcutta: Renaissance, 1955), p. 301.

From Outside the Imperium: Gandhi's Cultural Critique of the West

I

When one believes in an alternative vision of history, ... he is stepping outside the city to see a pastoral vision in which the office buildings and the universities do not obscure the archaic stars. All through history, from Abraham to Mao, prophets have left the city behind them to insist upon a vision of things greater than they are; but in the double nature of all phenomena, the abandoning of the city for the wilderness is also the pattern of madness: the psychotic leaves the social structure of sanity. From the psychotic's point of view one could paraphrase Voltaire to say that sanity is the lie commonly agreed upon. Those left behind in the city define themselves as responsible and sane and see the wanderer as a madman. The wanderer defines himself as the only sane person in a city of the insane and walks out in search of other possibilities. All history seems to pulse in this rhythm of urban views and pastoral visions.

William Thompson[1]

For a major critic of cultures in this century, Gandhi's sensitivity to many elements of culture and to many forms of ethnicity was oddly obtuse, if not altogether absent. If cultural sensitivity means awareness and acceptance of basic cultural differences, social anthropologists should straightaway disqualify Gandhi as one of their lay counterparts. Certainly he was no cultural relativist. In the thousands of pages of his collected works, there is hardly a sentence to suggest that he believed in fundamental or irreconcilable differences between cultures. And there is positive evidence that he put all his faith in universal, as distinct from cross-cultural forms of social theory. The assumption of universal values was so deeply ingrained in him

[1] William I. Thompson, *At the Edge of History: Speculations on the Transformation of Culture* (New York: Harper Colophone, 1972), pp. 214–15.

that civilizations, for him, always cut across conventional boundaries of cultures. As a result, the East and the West regularly met in his worldview. Stripped of its modern self, the Western civilization to him was not essentially different from its Eastern counterpart. When he talked of the differences between the Occident and the Orient, he mainly had polities and societies, not cultures, in mind.[2]

It is true that once, when asked 'What do you think of Western civilization?', Gandhi had sarcastically said, 'I think it would be a good idea.'[3] But it is also true that almost all his gurus were Western intellectuals. Even the two Indian intellectuals he believed to be his gurus were conspicuously bicultural: Rabindranath Tagore and Gopal Krishna Gokhale. The only book he read as much as the Gita was the Bible, and whereas his knowledge of the former was partly second-hand, that of the latter was first-hand. The closest friend he had in adulthood was an Englishman, C. F. Andrews, and it was an English intellectual, John Middleton Murry, who claimed that Gandhi was 'the greatest Christian teacher in the modern world'.[4] Even when choosing his political heir, Gandhi chose the Westernized Nehru rather than the more obviously Indian Vallabhbhai Patel and to the end his reference group remained the much-maligned Christian missionaries rather than the more Indian Hindu ascetics. Finally, by his own admission, he picked up his idea of non-violence not from the Hindu scriptures but from the Sermon on the Mount.

Gandhi's major criticism was directed against the modern West and its encroachment on the rest of the globe. He called this West the 'modern civilization' and lived with the hope that the other West would one day reassert itself, that out of

[2] When Gandhi spoke of pre- or non-modern cultures, he granted the existence of some cultural differences. But such cultures were seen by him as perfectly capable of intercultural dialogue. See his sixteen-point thesis quoted in T. K. Mahadevan, *Dvija* (New Delhi: Affiliated East-West Press, 1977), pp. 118–19.

[3] Quoted in T. S. Ananthu, 'Going Beyond the Intellect: A Gandhian Approach to Scientific Education' (New Delhi: Gandhi Peace Foundation, 1981, mimeographed), p. 1. See also Gandhi, quoted in Mahadevan, *Dvija*, pp. 151, 180.

[4] C. F. Andrews, *Mahatma Gandhi's Ideas* (New York: Macmillan, 1930), p. *192*.

the womb of the present would emerge *not* a single non-oppressive, egalitarian, urban-industrial, technocratic civilization but an authentic Western and an authentic Eastern civilization: non-oppressive, egalitarian but also primarily non-urban, non-industrial and non-technocratic. Gandhi stood against Western traditions only to the extent that the West had traditionalized modernity. He stood for the East only to the extent the East stood—by intent or by default—against the modern culture.

To understand this critique of the modern West—by this I mean the traditions and state of mind identified with the modern West, whether in the West or outside—one must discuss Gandhi along a number of dimensions, some of which may seem strange and esoteric. But all provide clues to the way a large part of the non-modern world has, for good or ill, understood, coped with or resisted the loving embrace of modernity. These dimensions do not exhaust Gandhi or provide a comprehensive Gandhian model of culture; they only represent *one* parsimonious analysis of Gandhi's ideas and *one* particular reading of his life. Gandhi's was not a systematic theory outlining an alternative, non-modern, non-Western society. His was a vision which triggered imageries of another class of society, latent in the minds of men living in, or with, modernity. His search was for what A. K. Saran, using a Platonic concept used previously by Ananda Coomaraswamy, calls metanoia, a situation in which more and more people are in their right minds. For that same reason, Gandhi was not *one* single critic of the modern West; he represented a whole class of critics of modern civilization. And like many others in the class he can be interpreted or reinterpreted in more than one way. To deduce one final supervening Gandhi from his life and work would be both anti-Gandhian and self-defeating.

II

Gandhi rejected the modern West primarily because of its secular scientific worldview. To him a culture which did not have a theory of transcendence could not be morally or cognitively acceptable. He knew that the ideological core of the modern world was post-Galilean science which prided itself on being the only fully secular area of knowledge. He also knew that legitimation of the modern West as a superior culture came from an ideology which viewed secularized societies as superior to the non-secular ones; once one accepted the ideology, the superiority of the West became an objective evaluation. One could then deny it only through intellectual subterfuges.

Gandhi's rejection of modern science is by far the best-known theme in his attack on the West and, for many, clear evidence of his tendency to compromise with orthodoxy and superstitions.[5] A few, however, have sensed that this aspect of Gandhi's critique is directed at the heart of the urban–industrial vision. They have found Gandhi uncompromisingly—often intolerably and irrationally—radical. They have recognized that, despite the half-hearted efforts of some like Albert Einstein, who have stood for forms of transcendence through science, modern science cannot but secularize larger and larger areas of our life. It has to do so because of the double 'split' in consciousness it induces. It promises 'true' knowledge, and the control and predictability which goes with such knowledge, only when a person (1) isolates or splits off his cognition from his feelings and ethics and (2) when he partitions himself off from the subjects of his

---

[5] Gandhi died, almost necessarily, at the hands of one who represented the modern world and sought a secular-scientific orientation to statecraft. The young assassin, Nathuram Vinayak Godse, supposedly a religious fanatic, gave a spirited last speech in court before the death sentence was passed on him. It was essentially a fervent, rationalist, modern plea to recognize the dangers Gandhi posed to the growth of the modern state in India and to the conduct of 'normal' politics along the lines Professor Henry Kissinger would have approved of. The plea invoked interesting reactions. Jawaharlal Nehru, for instance, called Godse an insane killer who did not know what he had done. Yet, Nehru's government banned Godse's last testament lest others should find it too sane. The government knew that it was Godse who was seeking the secular solution, Gandhi the religious. See Ashis Nandy, 'The Final Encounter: The Politics of the Assassination of Gandhi', in *At the Edge of Psychology: Essays in Politics and Culture* (New Delhi: Oxford University Press, 1980), pp. 70–98. The argument has been further developed in 'Godse Killed Gandhi?', *Resurgence*, January–February, 1983, (96), pp. 28–9.

enquiry emotionally. The two splits together constitute the kernel of the modern scientific worldview; experimentation is only an epistemic attempt to work it out.[6] Many elements of the modern life—for instance, the emphasis on a negotiable, market-oriented concept of equality and the totally instrumental, non-sacramental concept of nature—can be said to be the indirect expression of this aspect of modern science and its attempt to become universal by being non-ethnic, amoral and dispassionate.

Gandhi's response to the worldview of modern science was only partly Hindu. He did not take advantage of the classical concepts of knowledge in India to reconcile modern science with Brahmanic Hinduism. That response had already been tried out with great success by some of the social and religious reformers of nineteenth-century India. They had ensured that the Hindu did not have to fully secularize himself to be either scientific or rational; he only had to compartmentalize his life. Even the ultra-Hindus among nineteenth-century reformers, grappling with deep feelings of inferiority and trying to enter into a power game with Christianity and Islam, had sought to uphold purely secular and scientific considerations in public life and to demarcate religion and culture from politics and knowledge. Gandhi, on the other hand, recognized that mainstream Hinduism—the mix of the classical and the folk with which a majority of Indians live—had continued to be suspicious of the culture of modern science, as repeatedly shown by the pathetic appeals for more rationality and a more scientific spirit emanating from India's modern sector.[7] He wanted to use this suspicion as the basis of an Indian critique of modern secularism and modern science and of the new forms of

[6] On the splits in consciousness see 'Science, Authoritarianism and Culture' in this volume. Also, Ashis Nandy, 'Towards an Alternative Politics of Psychology', *International Social Science Journal*, 1983, *35*, pp. 323–38. The most penetrating analysis of Gandhi's attempt to desecularize public life is in A. K. Saran's 'Gandhi and the Concept of Politics: Towards a Normal Civilization', *Gandhi Marg*, 1980, *1*, pp. 675–726.

[7] For a recent example see P. N. Haksar et al., 'A Statement on Scientific Temper', *Mainstream*, 25 July 1981.

violence and oppression which went with them.

It is often said that the defeat of the church in the power-struggle between the church and science in Western Christendom lies at the root of the present-day Western concept of secularism and the cultural power of modern science and its version of rationality. Some have even argued that the struggle between the church and science was not that between obscurantism and reason, but between a narrow but ethically sensitive and plural concept of science as represented by the church and an ambitious but closed and amoral concept of knowledge as preached by the Galileans. Whatever be the truth of the matter, this old church-versus-science debate had become relevant in a different way in Gandhi's time. In a world where modern science had come to enjoy near-total hegemony and was in close alliance with the dominant powers, certain forms of 'irrationality' had become a defence—both in the psychoanalytic and the political sense—against totalization. Gandhi understood this. He knew that a tradition of coexistence between religion and science had grown in his society. India's experience with modern science and social reform had already shown that, in spite of the popular stereotype of the Indian society being organized around religion, Hinduism was the first social system to respond to—and legitimize—modern science in colonial India.[8] The options this opened up and the dangers this posed were probably sensed by Gandhi. Certainly it was already becoming clear in his time that by opening up new hierarchies, by linking up with the same aspects of Indian classicism, modern science was re-endorsing Brahmanism, which Gandhi was fighting, and devaluing the folk and the non-literate, for which Gandhi was trying to create a new space in Indian public life.

It is a measure of the success of modern science in India that his ultra-Hindu, Brahmanic assassin accused Gandhi of bringing in anti-scientific ideas like soul force and morality into politics. Nathuram Godse claimed that he had unwillingly to kill

---

[8] See Ashis Nandy, 'The Making and Unmaking of Political Cultures in India', *At the Edge of Psychology*, pp. 47–69.

Gandhi on behalf of the modern world, especially on behalf of modern ideas of statecraft and rationality, so that the newborn Indian nation could survive. One of Godse's last wishes was to take the appeal against the death sentence passed on him for killing Gandhi to the Privy Council in Britain—still the highest court of appeal for India in 1948—so that the world could judge his action impartially. He felt that the modern world would give him a better hearing than the superstitious, effeminate, Hindu admirers of Gandhi in India.

The moral of the story is this: Gandhi never accepted either modern science—or its extensions within the various secular theories of liberation—as the baseline for all social criticism. He saw religions and traditions themselves as a means of criticizing the existing and challenging the dominant. Like Karl Marx, he was aware that religion could faithfully represent the sigh of the oppressed and could be the heart of a heartless world. Unlike Marx, he sought to use this awareness as a weapon. He refused to separate politics from religion; one could separate the two, he felt, only if one accepted Machiavellism and realpolitik as the crux of politics and opted for unbridled instrumentalism leading to an artificial distinction between ends and means. This itself was a form of criticism of modernity. Realpolitik was not born in the modern period, but it was the modern world that gave it a new legitimacy as part of an applied science of politics. To defy such realpolitik and instrumentalism in politics was to defy both the economic modernity vended by the liberals and the theory of salvation which some radicals ventured as a critique of modernity.

E. F. Schumacher once drew attention to the following warning in the gospel according to St Keynes:

But beware. For at least another hundred years we must pretend to ourselves and to everyone that fair is foul and foul is fair; for foul is useful and fair is not. Avarice and usury and precaution must be our gods for a little longer still. For *only* they can lead us out of the tunnel of necessity into daylight.[9]

[9] Quoted in E. F. Schumacher, *Small is Beautiful* (London: Blood and Briggs, 1973), p. 20.

I have been told that the quote has a meaning entirely different from the one the overly suspicious non-economists derive from it. Perhaps. To me, however, it seems obvious that Keynes is only stating in a dramatic fashion an axiom which cuts across the various versions of liberalism, namely that the self-interest of individuals, when allowed to cumulate in the market-place of political economy, would produce collective good.

Unfortunately for those who swear by the moral instrumentalism of Professor Keynes, naughty theories do not look so clever after a few decades when time and human experience have caught up with them. For the amoral individualism Keynes recommends may bear, under conditions of scarce resources, an inverse relationship with collective good, thanks to the remorseless logic of what Garrett Hardin calls 'the tragedy of the commons'.[10] Under such conditions, the 'system' may ensure that the person's self-interest has a higher premium than the interest of the collectivity, but gives diminishing returns to every person in the system till tragedy strikes the collectivity.

Marx—or at least Marx collaborating with English positivism and scientism in the form of that eminent Victorian, Friedrich Engels—granted full legitimacy to modern science, placing secular consciousness at the centre of his utopia and freeing science from the constraints of history. Though there are major divides within the Marxist tradition about the extent to which dialectical materialism applies to nature, there is substantial consensus among Marxists that the social sciences are ideologically more tainted than the natural sciences. Moreover, but for a few, even the Marxists who plead for acceptance of the dialectics of nature use that acceptance as an escape clause—to free science from social and ethical constraints and to place it in the 'base' rather than in the 'superstructure'. Accept the truth of dialectical science, the line goes, and seek useful application of science on the basis of a correct view of its production

[10] Garrett Hardin, 'The Tragedy of the Commons', in Robin Clarke (ed.), *Notes for the Future: An Alternative History of the Past Decade* (New York: Universe, 1976), pp. 68–81.

relations, and you are on the right side of history.[11]

Rejecting both these traditions of social analysis, Gandhi denied the Baconian faith that the values of the Enlightenment could be actualized by the simple expedient of scientifically discovering and using secular trends in nature, society and history. He refused to believe that there could be an objective social knowledge independent of the knowers or an objective history impervious to the moral choice of individuals. Institutions could never be designed so perfectly or scientifically, he once pointed out, that they would obviate the need for individuals to be good.

The ideology of modern science, Gandhi felt, was endorsed by the ideology of modern technology. Technology had always had elements of instrumentality and control associated with it, even in the traditional societies. But modernity, by removing the older checks on instrumentalism, opens the way to technocracy. It derives from the secular scientific worldview a discourse in which the physical sciences provide the metaphor for the understanding and use of all living beings. A fully secularized, modern society cannot but produce a mechanomorphic concept of society and derive social priorities from it. To such a society the humanness of man is an embarrassment—exactly as the patient's understanding of his disease and his treatment is an embarrassment to the modern doctor. However committed to democracy in politics, such a doctor has to deny the relevance of democratic values in his profession. To him a silent patient is scientifically the ideal patient and a post-mortem the ideal context for medical objectivity.

Gandhi refused to grant total autonomy and control to the technologists. Many of his notorious whims derived from this refusal. Thus, he sought to recover the human body from the medical technologist and to give it back to the patient. He

---

[11] A good summary of the state of Marxian scholarship on this issue and its ethical implications is Diane B. Paul's 'Marxism, Darwinism, and the Theory of Two Sciences', *Marxist Perspectives*, Spring 1979, pp. 116–43. Also see Phil Slater (ed.), *Outlines of a Critique of Technology* (London: Inkluks, 1980); particularly the introduction by Monika Reinfelder, pp. 9-37.

refused to accept the doctor's authority over the body. Instead, he sought to restore the body to the individual by accepting the individual's right to define it. In his world, the patient's theory of health had primacy over that of the professionals.

I have separated scientism from technicism because, from the Gandhian point of view, the pathologies of the two are not the same. The former, in the name of fighting medievalism, promotes a form of hard materialism which negates even the idea of future freedom from material bondage which was Marx's dream. Such materialism becomes an end in itself. It reduces human rationality to a particularly narrow version of objectivity and objectification and it defines large parts of critical consciousness as irrational, romantic irrelevancies.

Technicism, on the other hand, further hierarchizes the relationship between human beings and nature and between those who possess technology and those who do not. It introduces a concept of social change which, in the name of science, rationality and social creativity, allows one to destroy parts of a person, society or nature for the latter's own good. The current euphemisms for such imposed salvation or social engineering are 'development', 'growth' or 'progress'. Such engineering in our times has been the prerogative not only of the colonial and neocolonial powers and their liberal detractors, but also of old-style revolutionary vanguards and their new incarnations in the 'conscientizers' of the underprivileged. Apparently, not only conformism but even dissent has been co-opted by the technocratic vision. As an early admirer of Gandhi, Lanza del Vasto, says, 'Man has been conquered. Twice conquered: convinced. He no longer protests, not even inwardly.'[12]

Technicism allows no escape, for the solutions to all technological crises are supposed to be in technology itself. As Peter Medawar, one of the cleverer ideologues of such technicism, declaims in the context of ecology: 'The deterioration of the environment produced by technology is a technological problem of which technology has found, is finding, and will

[12] Lanza del Vasto, *Return to the Source*, trans. Jean Sidgwick (New York: Simon and Schuster, 1971), p. 112.

continue to find solutions.'[13] He speaks as if the proportion of technologically-induced environmental problems for which modern technology had found solutions was shown to be on the rise! As if he himself, when making this oracular statement, were carefully basing himself on the empiricism he would otherwise swear by! It is such a tendency, to reduce all criticisms to internal criticisms, which is the major differentia of the ideology of modern technology. Gandhi's anti-technicism seems so devastatingly retrograde because it seeks solutions to the technological problems of our times partly outside technology. He refused to give in to the newspeak which declares all extra-technological attacks on technological problems to be 'utopian' or 'romantic' and all reformism—the faith that technological crises can be solved within the technological system—to be radical.

Gandhi rejected technicism, not technology. Had it been the other way round, he would have been found more palatable by many. Modernity knows how to deal with those who are anti-science or anti-technology; it does not know how to deal with those using plural concepts of science and technology. Here lay Gandhi's dissent. He was offensive and arrogant enough to offer another concept of technology. He was neither a mystic who believed in the absolute superiority of the spiritual over the material nor a romantic who believed in the superiority of the natural over the man-made. (One of the reasons he seemed a hypocrite to many was the shrewd this-worldliness with which he organized and ran his movements. A man of religion is not supposed to do so, according to the popular sociology of religion. That is why the modern Indian, too, prefers to posit hard political and secular choices behind the 'screen' of Gandhi's religious idiom and feels he has explained Gandhi by 'demystifying' the Gandhian talk of soul force, non-violence and self-punishment to reveal the 'real' choices underneath. To the moderns, scepticism is a purely modern prerogative and it must work in only one direction—it must show

---

[13] Peter Medawar, *The Hope of Progress: A Scientist Looks at Problems in Philosophy, Literature and Science* (Garden City, N.Y.: Anchor Doubleday, 1973), p. 135.

the 'base', in both senses of the word, beneath the superstructure of the good. Such scepticism can have no patience with a Gandhi who did not work with the standard dichotomy between spiritual 'superstructure' and material 'base', but unashamedly used both as the starting points for social criticism and for exercising suspicion.) He was a shrewd, sceptical, thisworldly *bania*, suspicious of all prophets, final answers and keys to history, indeed of all concepts of total good or ultimate evil. He *had* to reject the idea of a universal, cumulative, imperial technology, developing according to laws of linear progress.[14] Modern technology to him could not be the last word even in technology.

As a general rule, Gandhi was against technologies which replaced the uniquely human aspects of man. Not only because such replacement turned a person into a mechanical part of the production machine, but also because it turned him into a mechanical, 'dead' consumer of utilities. For Gandhi, technology had to be judged both on the grounds of what it did and what it symbolized.[15] In advocating the use of a machine like the *charkha* or spinning wheel, his argument would have been that the *charkha* was a morally superior and a techno-economically more effective machine than the cotton mill because (1) it did not supplant human beings; (2) it symbolized the dignity and autonomy of the individual resisting the demands of modern collectivities; and (3) it symbolized pre-modern technology and non-alienated labour. This in spite of the fact that the spinning wheel he popularized was a new edition of the traditional *charkha*. It was this 'odd' commitment to another technology which made him often work so hard on proper techno-

[14] Cf. J. D. Sethi, *Gandhi Today* (New Delhi: Vikas, 1978), pp. 30–1. Sethi seems to feel that Gandhi partly believed in the linear progress of science and technology. This obviously was not what Gandhi said in his sixteen-point thesis (Mahadevan, *Dvija*, pp. 118–19) as well as in *Hind Swaraj*. Sethi extrapolates, and that extrapolation is not consistent with my reading of Gandhi. Examples of machines Gandhi considered non-alienating, apart from the *charkha*, are the lathe, the bicycle and the sewing machine.

[15] Cf. Seyyed H. Nasr, *Man and Nature: The Spiritual Crisis of Modern Man* (London: Mandila, 1976).

logical problems as part of his political and social programme. He was neither a Luddite nor an unalloyed admirer of traditional technology. And he did reject many aspects of traditional technology in India as alienating, dehumanizing or as a plain simple bore.

The issue of modern technology is also the issue of modern technocracy—the rule of the experts who have the 'right' technology and the 'right' techniques and the rule of the organized sector from within which the experts operate. Perhaps Gandhi vaguely realized that modernity has splintered the popular concept of self-realization. The modern man no longer sees self-realization as a unitary end; he seeks it separately in as many spheres of life as possible. And this through the attainment of technical skills—in child-rearing, sports, work relations, anxiety-relief, friendship, bereavement, etc.—under the guidance of professionals or specialists.

Underlying this fragmentation of the ends of life lies a common personality factor. People seek, as Erich Fromm noticed, what they are programmed to seek. Whether it is Dr Adelle Davis speaking on our daily bread or the appropriately named Dr Alex Comfort comforting us on our daily bed, the modern worldview cannot but increasingly organize more and more of our lives. This organizational thrust has produced new kinds of stakes on which to burn a person to save his or her soul. We are fond of recounting, as a neat example of barbarism, that a few million women were done to death as witches in medieval Europe. We are less prone to recount similar cultural achievements in our times, the more so when such achievements come as part of a secular theory of salvation. Development, for instance, has become a new reason of State. Societies could impose today virtually any suffering on any number of their own people in the name of development, exactly as in earlier times, under the guidance of experts in matters of soul, witches were killed so that their souls could be saved. Simultaneously, the professional developmentalists have now acquired the right to define both the ideal person and the ideal community, as well as the right to legislate away all alternative ideals as so much heartlessness and retrogressive rhetoric.

Similarly, the idea of revolution, too, has become associated with professionalism and managerialism, leaving no scope for spontaneity, ambiguities, intuitions, innovations or poetry. The nineteenth-century optimism which sought through the positive sciences the liberation of humanity, had at its core a faith in professionalism or technicality. That faith has gradually taken over the socialization and education of modern man and woman, so that even when they defy authority, they have to look to another authority for professional guidance, and to set the standards of performance.

It was against this technicism that Gandhi rebelled. When serious Gandhians call him an anarchist, they have this rebellion vaguely in mind. Yet, strangely enough, this is one area where Gandhi's critique of modernity failed to take full measure of the problem. Operating in an anarchic culture which did not even have the tradition of an organized priestly class, he never identified the double-bind within which the fully socialized person in modern society is caught. When such a person is willing to conform to his repressive superego or conscience it becomes neurosis—the province of the psychiatrist, psychologist or psychoanalyst. But the refusal or inability to perform is also seen as a psychopathology—as deviance, alienation or self-destructiveness—which deserves to be handled by another set of experts. Either way, one becomes the ward of a professional. Either way, salvation, secular or otherwise, ceases to be a personal search. It is enjoined by a priesthood, a vanguard, or a technocrat. In other words salvation is sought under the guidance of social engineers to whom the arithmetic of pain (and the painful pleasure of calculating how many must suffer for the sake of how many and how much one must suffer for how much gain) is the central issue in any suffering.[16]

[16] On the exteriorization and objectification of pain under modernity, see Ivan Illich, *Medical Nemesis: The Expropriation of Health* (New York: Pantheon, 1976). Also Liam Hudson, *Human Beings: An Introduction to the Psychology of Experience* (Fragmore, U.K.: Triad Paladin, 1978), ch. 13. Hudson, responding to the social criticism implied in R. D. Laing's ontology of madness and sanity, posits two

Even though he underestimated the cultural power of the modern thought-machine, it is partly as an indirect result of Gandhi's attempt to bring some of these issues into the public realm that a number of scholars have recently spoken of the dehumanization brought about by the deadly combination of technicism, expertise and over-organization.[17] It appears from these works that such dehumanization is neither an accidental by-product of modernity nor an aberration from it. It is inherent in the logic of what has been called the age of professions. It is a part of the principle of rationalization which Max Weber idealized on behalf of our times and which, according to Jürgen Habermas, frequently becomes rationalization in the psychoanalytic sense. It is in the spirit of technocratic revolution to produce, as Hannah Arendt recognized, the banal evil or bureaucratic violence which shapes so much of public life in our times.

Elsewhere I have discussed Gandhi's critique of hyper-adulthood and hyper-masculinity.[18] Here I shall only touch upon the issue to underline the linkage in the popular culture of modernity between the ideology of adulthood and masculinity, on the one hand, and the ideology of modern science and technology, on the other. (This is one more area where metaphors of gender and age determine the social status and the truth value of knowledge systems.)

True adulthood, the modern culture claims, can be acquired

---

forms of rationality: ends–means rationality and intrinsic rationality. He defines sanity as a balance between the two forms. Perhaps occasionally polarization takes place between those who think sanity demands more ends–means rationality (presumably similar to instrumental rationality in critical theory) and those who think that sanity demands less ends–means rationality. Gandhi was sure that madness in his times meant, in Hudson's words, 'the tyranny of sensibility over sense' (pp. 171–4) and the psychopathy induced by overemphasis on means–ends rationality.

[17] For example, Ivan Illich, *Deschooling Society* (New York: Harper and Row, 1971); *Medical Nemesis*; and *Tools for Conviviality* (New York: Harper and Row, 1973).

[18] Ashis Nandy, *The Intimate Enemy: Loss and Recovery of Self under Colonialism* (New Delhi: Oxford University Press, 1983). I have also argued in 'The Final Encounter' that Gandhi's death was at the hands of an ascetic young man

through scientific rationality, restraining the emotions and one's indisciplined, non-rational, speculative self. True masculinity is the hard-headed, performance-minded pragmatism of technology. The older civilizations and the primitives, the argument goes, lack the higher rationality of modern science, just as the intuition-laden, effeminate laity lack the technologist's tough-mindedness. The professional scientist is adult because he has impersonal, dispassionate, sanitized objectivity; the layman can only hope to be mature by imitating him. The professional technologist is masculine because he is tough, practical and value-neutral; non-technologists can only hope to be potent by being technique-minded in their daily life and by employing or consulting the technologist.[19]

Thus the crucial role of the ideologies of adulthood and masculinity in modern oppression, particularly in colonial encounters between the East and the West in Asia and Africa. When in the second half of the nineteenth century, power shifted from the feudal to the middle classes amongst the European colonists, especially in India where the subtler outlines of the colonizer's theory of colonialism were first drawn—the new power elite sought in modern science and technology justification of the white man's immense civilizational burden. The hierarchical principles deriving from age and sex, already prominent in European cosmology, were combined with this justification to forge a composite picture of the colonizer—scientific, technologically skilled, bearing the responsibility of history and dutifully playing a civilizing role. By definition the colonized now became wilful, negativistic, reprobate, and passively aggressive like a child or a woman.

It was Gandhi who first systematically defied these homo-

from a 'martial' background who thought Gandhi a menace to the forces which were trying to reclaim for India (the Hindus) its lost virility and its status as an adult and masculine competitor with Christianity and Islam.

[19] The reverse view was reflected in some strands of consciousness in the traditional West which saw the man not dependent on technology as more masculine. In the modern West, however, science and technology are clearly seen as masculine pursuits, even though studies of scientists and technologies suggest that creativity

logies or identities with his concepts of adulthood and maleness. Sensitized by the immense 'wealth' of colonial literature
produced by the Kiplings and the Stracheys, and even by Marx
and the two Mills—and of course, by his own early experience
as a tortured, youthful nationalist, seeking Kshatriya potency[20]
—Gandhi sought a more inclusive maleness which would include femininity, particularly maternity, and a more inclusive
adulthood which would not shy away from what Kipling called
the half-savage-half-child self of the colonized. In a civilization
in which the gods still displayed androgynous qualities, such
concepts of masculinity and adulthood not merely defied the
colonial culture but reaffirmed indigenous categories and indigenous visions of a desirable society.

There is an apparent contradiction here. If age and sex are
used as principles of hierarchy, they do provide scope for treat-
ing the primitive as a person, even if as a childish or a feminine
one. How do such hierarchies fit in with the objectification
under colonialism which Aimé Césaire depicts or with the de-
sacralization under modernity which Ananda Coomaraswamy
describes?

One answer is that colonialism modernized and adapted for
its own use the traditional Western hierarchy, extending from
gods to inanimate objects:

gods > humans (adult men > women or children) > animals > things

In the process colonialism reactivated the fear of liminality
which women and children invoked in the European culture
by being at the margin between human beings and nature.
Previously, this fear had accompanied some fear of the power
of nature and of the sacredness and magicality associated with
nature. These set limits on objectification. Under colonialism,
as under industrialism, the secular, modern worldview removed

---

in the area is closely associated with femininity. Compare for instance Brian
Easlea, *Fathering the Unthinkable* (London: Pluto, 1983) and Anne Roe, *The
Making of a Scientist* (New York: Dodd, Mead, 1953).

[20] Erik H. Erikson, *Gandhi's Truth: On the Origins of Militant Nonviolence* (New
York: Norton, 1969).

these limits. There was now fear of being feminine but no fear of the feminine and certainly no fear of falling foul of the feminine principle in the cosmos. It was in this context that Gandhi's androgyny sought to give back to femininity a part of the traditional sacredness and magic associated with it. In an attempt to de-objectify the colonized, he tried to de-objectify women, too.

The flip-side of Gandhi's resistance to the modern use of the metaphors of gender and age was his implied criticism of normality and over-socialization. All his life Gandhi suffered from the image of being a political eccentric. Winston Churchill's epithet, 'a half-naked fakir', and his objection to Gandhi going in his loin-cloth to negotiate on equal terms with the British Viceroy in India was more than a racist slur; it was an attempt to banish Gandhi from normal politics. It was along the same continuum that George Bernard Shaw said, after Gandhi's death, that his death only showed how dangerous it was to be too good. It was a Fabian attempt to place Gandhi as a visionary outside history and outside politics. Both men, in assessing Gandhi, were using the dominant Western concept of normality. Churchill was referring to its everyday version when he said that he found 'nauseating' and 'humiliating' Gandhi's dress; he was expressing the belief of his class that the formally-attired person was the only attired person, for sartorial respectability linked up normal politics with other forms of normality for the British gentry.[21] (Gandhi, on his part, could be bitterly sarcastic about this linkage. Once when asked about his scanty clothes after a visit to Buckingham Palace, his remark was 'The King had enough on for both of us.') Shaw of course was referring to that other version of normality in which normal politics was 'true' or unmasked politics or, it comes to the same thing, 'dirty politics', realpolitik or Machiavellism.

Both Churchill and Shaw recognized that Gandhi was defying not merely political authorities but also authoritative political

[21] Cf. Ali Mazrui, 'The Robes of Rebellion: Sex, Dress and Politics in Africa', *Encounter*, 1970, *33*, pp. 19–30.

myths. Both were aware that, unlike the liberal and socialist thinkers, Gandhi was attacking the basic elements of the culture which held together the modern West. These elements were (1) the image of normal politics as a non-synergic game in which each person's gain is another person's loss; (2) the idea that the normal politics of self-interest, if properly managed, contributes to the social good and to humane social arrangements; what Gandhi called the vision of a social structure so perfect that it would obviate the need for personal goodness; (3) the separation of normal politics from the search for self-realization, as the metaphor of self-realization is seen as a dangerous mystification which justifies oriental despotism and delinks a person's actions from the market; and (4) the belief, underlying much of the modern critical tradition, that the 'dirtier' and the crueller is always the more real self of man and that all altruism is, in the final analysis, a social imposition or cultural artefact.

Tacitly, both Churchill and Shaw were referring to Gandhi's madness, to his 'transcendent eccentricity' as A. K. Saran euphemizes it. As we well know, in modern society moral judgements do not apply to the person who is mad; they apply only to his actions. Ignorance of law is no defence under Anglo-Saxon law; but madness is. Under such law, the primitive can avoid being adjudged guilty only if he accepts the Freudian equation between primitivism and madness. Madness thus is not a purely negative epithet in the post-medieval West; it is also a means of making irrelevant a person whom the West's moral self cannot ignore but its practical self would like to reject. Hence the new equation between madness, primitivism and infantility is supposed to arouse, in the sane modern adult, simultaneous feelings of hostility, nurture, and responsibility of the kind which goes with firm, expert 'management'. And for those subject to the power of such sane adults, survival means accepting, at least overtly, the modern definitions of madness and sanity. I need hardly add that this was what the colonial theory of progress was all about.

For Gandhi on the other hand, as for R. D. Laing when he

speaks of schizophrenia, breakthroughs could come through breakdowns. Normal civilization to Gandhi was not the conventionality produced by modern education and centralized mass media, operating as agents of uniformity. Normal civilization for him was one in which the dialectics between socialization and under-socialization—and between sanity and insanity —was never lost. Thus the Nietzschean paradoxes—admirers of Herbert Marcuse would say negations—of Gandhi: it is better education not to be educated in the modern times; it is more civil not to be civilized in the modern sense.[22]

Much of Gandhi's 'insanity' came from his hostility to nineteenth-century social evolutionism.[23] And nowhere was this hostility more apparent than in his rejection of the ideas of history and progress.

In the Western intellectual tradition there have been for long two opposite tendencies. One converts space into time: geocultural differences into historical stages and distant cultures into utopias or, as it sometimes happens, anti-utopias or dystopias. The other translates time into space: especially history into myths, and stages into types. A good example of the former is the belief of evolutionist anthropology that primitive societies represent the past of the modern West, and the psychoanalytic faith that there is an infantile primitive self of man not merely metaphorically here and now, but also diachronically and objectively in the child and the savage. The latent message here is that the future of the child and the savage is known because it is no other than the present of the adult and the civilized. An example of the second tendency is the latent psychoanalytic —and for that matter, medical—concept of history as case-

---

[22] Hence also T. K. Mahadevan's aphorism that Gandhi was 'so poorly read that his very ignorance was his strength' ('Mahatma Gandhi', *Youth Times*, 16–31 August 1980, pp. 14–15). Gandhi, it might appear, was keen to validate the inverse relationship between formal scholarship and creativity which has been shown in a number of psychological studies of the creative person.

[23] For a comprehensive statement on the anti-evolutionist stance of Gandhi, see Saran, 'Gandhi and the Concept of Politics'. Also Nandy, 'The Psychology of Colonialism'.

history, shaping as well as reflecting the immediate reality of the therapeutic situation and, thus, allowing one to 'work through' history. The patient can be healed, according to this concept, when the history of his illness is bared and the possibility of intervention in that historical process is opened up by this historical knowledge.

Though the second tendency has sometimes been popular among romantic counter-modernist movements and radical social critics, it is the first tendency, the one which converts space into time, that has come to dominate modern Western consciousness and to legitimize imperialism, the pathologies of modern science and the violence which accompanies development. Hence Gandhi's partiality for the second strand of consciousness, which also happened to have an important place in many traditional theories of time. For this second strand of consciousness all histories are contemporary myths and all myths camouflaged experience. For Gandhi, who preferred to be guided by myths rather than by history, by *puranas* rather than by *itihasa*, neither space nor time was or could be a pure category.

Critics of objectification have not often noticed that the subjects of 'scientific history' are subjects irrevocably and permanently. There can be some counter-transference (in the Freudian sense), empathy or partiality for his subjects in the historian; there can be no transference, no real dialogue, perceived mutuality or continuity between the historian and his subjects from the subjects' point of view. The historian's subjects are, after all, mostly dead. (This is unlike the subjects of the anthropologist who can show some semblance of defiance or rebellion against the anthropologist's construction of a culture by outliving him and appealing to the next generation of anthropologists. So much so that some anthropologists can hope that the salvation of anthropology—and that of the anthropologist—may well lie in the subjects of anthropology.[24])

One wonders if some vague awareness of this asymmetry

[24] T. N. Madan, 'Anthropology as the Mutual Interpretations of Culture: Indian Perspectives' (mimeographed).

between the subjects and the objects, and between the knowers
and the known, prompted Gandhi to reject history as a guide
to moral action and derive such guidance from his reading of
texts and myths. Surely he rejected evolutionism and the idea
of stages of growth, and he thought transformative politics and
social intervention possible without social engineering and
without what in the non-Western world has become an increasingly oppressive theory of progress. He would have agreed
with his Western follower, del Vasto, that 'men are carried
along now, not by the initial impulse of their faith in progress,
but by its continuing course and submission to fatality. . . .
They tell you "you can't go against the laws of history and
economy. There is no turning back now!"'[25] Gandhi, like
Blake and Thoreau before him, defied this new fatalism of our
times.

I have discussed Gandhi in terms of six themes which scaffolded
his critique of the modern West. This is not the way he would
have summed up his case, but this is the way he might have
summed it up for those living through the last two decades
of the twentieth century. In consonance with this approach,
I must now try to complete the picture by relating the Gandhian critique of the West to the problems of the mass society.

Gandhi did not live in a mass society nor did he work in a
country which showed portents of becoming a mass society. His
main reference points were the traditional Indian village and
its counterpoint in the pre-war Western city. Yet, either by
coincidence or because he intuitively sensed the inner logic of
nineteenth-century modernism, he stumbled on two main cultural features of the mass society in the advanced urban–industrial world. One of these has been often discussed under the
name of possessive individualism, a form of individualism
which, whatever might have been its scope and limits under
early capitalism, can reduce the individual in a mass society
to an active consumer of an expanding range of 'utilities', and
to a passive consumer of messages emanating from the mass

[25] del Vasto, *Return to the Source*, p. 110.

media. The early-capitalist ideal of the rugged individual as a producer and rightful consumer of the fruits of his toil becomes here the late-capitalist, mass-society ideal of the individual as a dutiful consumer, with an exponentially growing range of needs induced or sanctioned by the media.

Against these ideals Gandhi pitted an anarchic individualism which emphasized personal salvation, lonely dissent and the moral power of the individual *satyagrahi*. The popular belief is that in an unorganized society Gandhi had to stress collective action as a means of politicization and nation-building. He therefore emphasized sociality at the expense of individuality. It is, however, arguable that Gandhi rejected the split between the personal and the social; that he tried to base all collective action on a form of individual conscience which at its best had to represent the collective conscience rather than be related to the collective good through the individual's adjustment to the collectivity; that like his spiritual fellow-traveller Thoreau, Gandhi believed that a person more right than his neighbours was in a majority of one, for the individual's moral choice, when rightly arrived at and rightly exercised, was·for him a superior collective choice by definition. Such an approach to the collectivity could not be co-opted by the ideology of the mass society.

The other element of a mass society Gandhi anticipated, namely the presently popular idea of equality, needs to be examined in greater detail. On the face of it, Gandhi rejected equality. His concept of trusteeship in economics was certainly not egalitarian in the modern sense. Nor did his political style or the management of his *ashramas* reflect any egalitarian thrust as it is commonly understood. Yet, paradoxically, he built his nationalist movement by aggressively using and promoting some forms of equality, based on certain forms of individualism. Many accounts are available of how, while organizing the Indian freedom struggle, he set up sons and daughters against parents, individuals against their castes, and even his own personal political beliefs against the collective decisions of the organizations he led. And he did this not in the name of re-structuring traditions but in the name of strengthening Indian-

ness, Hinduism and *dharma*, all of which are supposedly collec-
tivist–hierarchical in orientation. Evidently, Gandhi attached
some sanctity to the defiance of unequal orders and hierarchies,
even if, as in the case of his movement against Brahmanic
Hinduism, he did not seemingly challenge the 'ancient' ideological basis of these hierarchies. (Probably because the hierarchies represented a traditional order being encroached upon by
the modern world at a time when the main civilizational problem for him was the encroachment; probably because he felt
that in the older order there were more checks on the collective
and the hierarchical than in the modern one.)

Gandhi's critical attitude to the modern idea of equality has
given contradictory messages to his interpreters. Thus, A. K.
Saran believes Gandhi upheld the traditional principle of hierarchy; others say he slipped from Hindu orthodoxy to Hindu
obscurantism when he embraced the idea of a *varna* system of
dehierarchized, purely occupation-based castes; still others
think that he subscribed to a vague egalitarianism but it was
a strategic compromise to mobilize his people for the nationalist
cause.[26] To make sense of these diverse interpretations, one must
recognize that Gandhi's concept—or alleged rejection—of
equality cannot be separated from his concepts of responsibility
and compassion, which in turn were, in his framework, extensions of his concept of *ahimsa* or non-violence. Gandhi sought
to epitomize in his life, not always successfully, a form of traditional social relationship which was not coloured by patriarchy,
authoritarianism or institutionalized violence, and which had
some degree of negotiability and self-criticism built into it.
This openness he wanted to ensure without recourse to the
contractual, impersonal, competitive individualism and the
part-object interpersonal relations on which a mass society
depends.

Hence, Gandhi respected the equality which was free from

[26] Saran, 'Gandhi and the Concept of Politics'. The last two positions are
both well exemplified by M. N. Roy's rationalist critique of Gandhi; see his
'M. K. Gandhi', in *Men I Met* (Bombay: Lalvani, 1968), pp. 26–31. See also
Jawaharlal Nehru, *An Autobiography* (London: The Bodley Head, 1936), for the
same ambivalence in a muted form.

the cultural assumptions of the modern society and the modern
market. Responsible, non-violent, non-contractual, non-com-
petitive, non-hegemonic, Buberian equality had a place in
Gandhi's life and in his theory of life. Equality for him, I like
to believe, was ultimately the mutuality and the nurture one
could provide another, not in return for or in exchange of
what the other had done for one but in 'equitable return' for
the over-all nurture one had received from the human and
non-human cosmos. This is a nurture which may or may not
have anything to do with the actions or performance of the
immediate target of nurture. I should perhaps add that Marx
at his best was always aware of a secular version of this concept
of equality. This is evident not only in the slogan about need
and ability that he popularized but also in the concepts of
what could be called 'non-economized equality' and 'sanctity
of individualized needs' in his *German Ideology* and *Critique of
the Gotha Programme.*[27]

III

In modern philosophy things are either so or not so; in eternal
philosophy this depends upon our point of view.
                                            A. K. Coomaraswamy[28]

About Gandhi's attitude to the modern West three questions
still remain to be answered. First, how correct was Gandhi in
his reading of the West? Can a clear line be drawn between
the Western and the modern civilizations? Many feel that such
a line cannot be drawn, even though they do admit a disjunction between the modern and the pre-modern West.[29] Modernity, according to them, follows from Westernness. While the
modern West was built on the ruins of the medieval West, was

[27] Karl Marx, *German Ideology* (Moscow: Progress Publishers, 1976), p. 566;
and *Critique of the Gotha Programme* (Moscow: Progress Publishers, 1978), pp. 11–18.

[28] Ananda K. Coomaraswamy, 'The Vedanta and Western Tradition', in
Selected Papers, ed. Roger Lipsey (Princeton, N.J.: Princeton University Press,
1977), Vol. 2, pp. 3–22; see p. 6.

[29] For a recent example, see Johan Galtung, Tore Heiestad, and Eric Ruge,
'On the Decline and Fall of Empires: The Roman Empire and Western Imperialism Compared' (Tokyo: United Nations University, mimeographed).

not the renaissance also a reaffirmation of some aspects of the classical West? What about the rekindled interest in ancient Greece and in the Hellenic culture after the renaissance? Is not the modern Western man—the Faustian man, as some prefer to call him—a reincarnation, a subspecies or a new cycle of the Promethean man?

At one plane, those who raise such questions are, on this issue at least, truer to Gandhian morals than Gandhi himself. Gandhi always took the responsibility for the unintended consequences of his political ideas. For him, there was nothing like accidental vulgarization of an idea system; all the possible vulgarizations of a system were inherent in the system, and its proponents and users had to own up to these possibilities as a personal responsibility. And the same principle applied to cultures. Thus Hinduism, according to him, could not disown the practice of untouchability. Nor could Gandhism disown responsibility for whatever violence it produced in practice.[30] Similarly, the West too, Gandhian norms would insist, should own up its oppressive self and the cultural continuity of that self with the West's classical and medieval selves. It cannot disown its present merely as an aberration from the true spirit of the Western civilization. On this plane, Gandhi's admiration for the pre-modern or non-modern West was not compatible with his ideas of cultural criticism and cultural responsibility.

However, Gandhi also believed in presuming an irreducible element of humanness in any antagonist or culture that was a part of his non-self.[31] He lived and worked *as if* that human-

---

[30] The angry reactions of many Indian nationalists when Gandhi called off some of his movements because they turned violent or when he went on fasts to expiate for such violence are well known. Such actions of Gandhi were not merely efforts to maintain the purity of means but efforts to maintain the moral links between theory and practice, science and technology, and between politics and society. He refused to accept the 'commonsense' proposition that a faulty vision could produce right action or a faulty action could follow from a correct vision. (For the moment I am ignoring, important though it is, the tremendous political timing which the Mahatma brought into his moral actions.)

[31] 'I myself have always believed in the honesty of my enemies', Gandhi once said. 'And if one believes in it hard enough, one finds it.' del Vasto, *Return to the Sources*, p. 123.

ness was always waiting to be rediscovered. Not being a social anthropologist, cross-cultural empathy or etic-emic differences had little significance for him. For him, an antagonist demanded ethical and cognitive respect on the basis of universal values, not on the basis of ethnic values peculiar to the antagonist. If necessary, one had to participate in the antagonist's painful struggle to identify his/her/its recessive self.

Thus, the Europe Gandhi accepted was not the Europe 'history' had known. But then Gandhi explicitly rejected history as a guide to moral action. He would probably have said that it was for the West and Western theorists of progress to admit to what they had done both to themselves and to the rest of humanity. The non-West for its part must act as if the last two hundred years had been a vulgarization of European civilization and Christianity.

How far was Gandhi's response to the modern West the response of a traditional Indian? Elsewhere I have described Gandhi's ideology as critical traditionalism but it was probably more complex than that, containing as it did a basic paradox. It is in the nature of traditional India to maintain a certain openness of cultural boundaries, a permeability which allows new influences to flow in and be integrated as a new set of age-old traditions—one may call the process traditionalization— and for some cultural elements to flow out and be detraditionalized. These two processes of inflow and outflow determine, at a given point of time, Indian culture, rather than a rigidly defined set of practices or products surviving from the society's past. Any attempt to produce or live by such a set has always gone against the spirit of Indian culture. Norms in India, *dharma*, have always been spatially and temporally specific. That is the first differentia of Indian culture and the main reason why the culture does not have a concrete concept of evil as does the Judaeo-Christian tradition. This of course also means that India, to be India, has to have the courage to be non-Indian—it is part of her *dharma* to be a microcosm of the world.

The role of 'marginality' therefore becomes doubly important for social creativity and cultural survival. Gandhi recognized this:

Everyone of the Indians who has achieved anything worth mentioning in any direction is the fruit, directly or indirectly, of western education. At the same time, whatever reaction for the better he may have had upon the people at large was due to the extent he retained his eastern culture.[32]

Over-definition is seen here as the first enemy of Indian tradition and creativity within that tradition, and bicultural sensitivities, when used within the native cultural frame, as a significant social resource. (A telling example of the destruction of a creative tradition through over-definition is the case of Hindu law which, once it was formalized in the British period, quickly froze at the level of the *dharmashastra* laws, freed from the changing demands of local and customary laws. This killed the vitality of the traditional law which thrived on the dialectic between the formal and the informal, the Brahmanic and the non-Brahmanic, and the cross-regional and the regional.)

Gandhi lived with an awareness of this dynamic. When he defended Hinduism he did not defend a religion or a theology. He defended an open-ended way of life. His Hinduism included many recognizably Western elements which he saw as necessary for contemporary Hinduism. The fluidity of the native concept of time helped him to see these elements as part of traditional Hinduism, in fact as part of a truer version of Hindu orthodoxy. Thus, Gandhi can be seen as very Indian if one goes by his attitudes to history and culture; he can be seen as very un-Indian if one goes by some of the specific items in his ideology and practice.

As telling is the paradox that, while Gandhi opposed over-organization and defied the age of professionalism, he was one of the most successful political organizers of our times. He organized the 'unorganizable' on issues which did not often seem good bases for mass mobilization. Those who were taken in by his ascetic idiom were often shocked by the this-worldly

[32] Quoted in Mahadevan, *Dvija*, pp. 178–9.

shrewdness and political acumen with which he responded to the organizational demands of mass politics. The concept of cultural relativism cannot explain this 'inconsistency'. Nor can the everyday concept of hypocrisy. First of all, there was, for Gandhi, no inconsistency. He seemed to believe that if the over-organization of the modern world could be criticized from the vantage point of the under-socialized aspects of man's social self, the overly formalized parts of Indian traditions, too, could be opened up with the help of tough politics seen as a recessive part of the Indian tradition. So he brought into organized mass politics face-to-face contacts and a private discourse emphasizing salvation, the inner voice, self-purification and truth. At the same time, he ruthlessly fought through organized politics the ideals of ritual purity and superiority of intellection in Brahmanic culture, the vulgarization of whatever might once have been positive about the principle of hierarchy in Indian public life, and the presumed irrelevance of politics to everyday life and personal salvation.

The creative possibilities of the approach were best revealed when Gandhi declared the armed Polish resistance against Nazi Germany in the late thirties to be non-violent. However dishonest such a statement from a man of peace may sound to modern ears, it underscored the difference between a world-view which sees violence as the organizing principle of political life (and non-violence as a sometimes-accidentally-successful political technique and as an exception to the rule) and a world-view which sees non-violence as the basic principle of politics within which the subsidiary principle of unavoidable violence has to be accommodated.

The issue can be posed in another way. Was Gandhi truly in the tradition of the great Hindu religious leaders who had, from time to time, interpreted and reinterpreted Hinduism in response to the changing world? Or was he in the tradition of those great nineteenth-century Indians who coped with the West by seeking support and inspiration in the native texts for new ways of thinking and living?

As far as the first question is concerned, Agehananda Bharati

gives the most honest answer. Gandhi, he politely says, was
ill-acquainted with Indian traditions.[33] By this Bharati of course
means that Gandhi neither knew the sacred texts well nor
knew the sophisticated commentaries on them, nor did he know
the weightages traditionally given to the different texts. Less
polemically, D. M. Datta, too, refers to the highly idiosyn-
cratic, 'non-technical', private meaning Gandhi gave to some
of the key concepts of Indian philosophy.[34] Neither Bharati nor
Datta seems aware that Gandhi had become, in his own life-
time, not only 'a great leader of the Hindus', as Mohammad
Ali Jinnah once said in back-handed homage, but also a religious reformer who had come back to the 'city' from the 'wilderness' with *his own* traditions of Hinduism.[35] He had acquired
the right to 'distort' authoritatively.[36]

Gandhi's Hinduism differed from that of some of his predecessors for the very reason Bharati disapproves: it reaffirmed
the non-canonical and the folk, on the assumption that, with
such a base, Indians would cope better with modernity. It was
by using the Vedas and the Upanishads, the *shrutis*, that the
earlier Indian reformers had grappled with the modern impact

[33] Agehananda Bharati, 'Gandhi's Interpretation of the Gita: An Anthropo-
logical Analysis', in Sibnarayan Ray (ed.), *Gandhi, India and the World: An
International Symposium* (Philadelphia: Temple University Press, 1970), pp. 57–70.
Cf. Mahadevan, 'Mahatma Gandhi', p. 15, who makes the subtler point that
Gandhi, if he had known Indian traditions more deeply, would have done better
and depended less on Western intellectuals like Ruskin, Thoreau and Tolstoy.

[34] Dhirendra Mohan Datta, *The Philosophy of Mahatma Gandhi* (Calcutta: University of Calcutta, 1968), especially pp. 23–32. See also note 24 above.

[35] Perhaps for some, Jinnah's homage had another unintended meaning: it
recognized that while the Hindu, Islamic, Buddhist, Sikh and Christian traditions are different in India, they use a common grammar. Thus it is possible to
be a leader of the Hindus while being a leader of India.

[36] This was Gandhi's strength as a political leader. His mobilizational capacities grew out of his ability to become simultaneously a political and a religious
leader. Bharati ('Gandhi's Interpretation of the Gita') himself gives a clue to
the nature of this particular form of charisma when he admits that even Shankara
in the eighth century erred by imposing a strictly monistic interpretation on the
Gita. To the extent the 'error' has dominated Indian consciousness for eleven
hundred years, one is entitled to ask why it cannot be read as a culture's own
style of interpreting its texts. A culture is not obliged to follow the principles
of good scholarship proposed by contemporary intellectuals.

of the West.[37] Gandhi joined those who used less catholic texts, the *smritis*, to take a more critical stance towards modernity. The Gita for instance was one non-canonical text which was constantly being sought to be made canonical by Indian social reformers from about the mid-nineteenth century. It gave less 'play' in the matter of reinterpretation than, say, the more metaphysical Upanishads. But it allowed a more politically activist stance, a better defence of everyday Hinduism, and a clearer link between the Brahmanic and the folk. It allowed one to break out of the overly modern interpretations of the Upanishads and the Vedas, offered by the likes of Rammohun Roy (1772-1833) earlier. It was through Gandhi that the Gita came closest to being a canonical text in Hindu consciousness.

Even Gandhi's relationship with the earlier social reformers of India—or their memories—had its in-built politics. The nature of that politics should be obvious from the impact he made and the ill-will he invited from the Brahmanic elites, particularly in Bengal, Maharashtra and south India. It was they who had led, during the previous hundred years, the society's struggle to cope with the modern West by giving new meanings to old categories. They had specialized knowledge of the Hindu texts and had used the knowledge effectively to adapt Hinduism to modern consciousness, often painlessly. But time had begun to pass them by even before Gandhi's entry into Indian politics. With the quickening pace and growing power of modernity, their efforts had become inconsistent with the self-esteem of a majority of the Indians being exposed to the modern sector. Modernity no longer seemed manageable within Indian traditions; it was threatening to take over the entire culture. The challenge now lay either in resisting modernity or in adapting it to Indian needs while denying it any intrinsic sanctity. The altered psychological environment demanded a shift in emphasis from a pure critique of traditions to a critique of traditions coupled to a critique of modernity. Gandhi represented this shift. Hence the pained tone of a liberal contemporary of Gandhi like Sankaran Nair who said: 'There is

[37] See Nandy, 'The Making and Unmaking of Political Culture in India'.

12

scarcely any item in Gandhi's programme which is not a complete violation of everything preached by the foremost sons of India till 1919.'[38] Nair sensed that Gandhi was not the culmination of the nineteenth-century Indian 'renaissance'; he represented a disjunctive experience of the society and a new critical tradition within Hinduism.

Gandhi was not, even in his time, the only major critic of the Western culture. He had predecessors as well as contemporaries. Some of them even ventured their criticisms roughly along his lines. Where then lay Gandhi's distinctiveness? And why has he risen phoenix-like from his ashes to haunt the 'rational', the 'normal' and the 'mature'?

The answer can be given at many levels. For instance, it is obvious that, unlike Rousseau or Freud, Gandhi did not consider civilization an incurable disease.[39] No tragic vision haunted his theory of salvation and he *did* think a 'normal civilization' possible. Probably for this very reason, he was willing to push his critique of modernity to its logical conclusion, unfettered by the fear of reaching a cultural dead end.

Again, of all the major critics of modernity, Gandhi was one of the few to offer a radical critique of urban-industrialism and modern science. And this without opting out of organized politics like a mystic or a saint. He would not accept the urban-industrial vision in the name of progress, and he refused to place science outside culture or history. Unlike Marx he did not seek to reform the social relationships of modernity; he rejected modernity itself. Unlike Mao Zedong, who shared some of his concerns, Gandhi never dreamt of entering a race with the modern West to beat it at its own game; he sensed the exhaustion of this civilization after four hundred years of Western exposure and two hundred years of colonialism. He envisioned a new game drawing upon some very old rules and conventions (as modernity itself had done when it bypassed the medieval experience to go back to the classical West). And

---

[38] Quoted in Mahadevan, *Dvija*, p. 134.
[39] Ibid., p. 14; also Saran, 'Gandhi and the Concept of Politics'.

unlike Freud, who while providing a fundamental critique of
the Western culture in Gandhi's time, was unaware of the
idealization of adulthood, masculinity and normality in his
own work, Gandhi was willing to be irresponsible, effeminate,
immature, and insane.[40]

It was this radicalism which linked Gandhi's ethics to his
cognition. Gandhi's mid-Victorian puritanism and almost
panicky fear of sexuality have popularized the idea that he
primarily rejected the ethics of modernity and the tinsel glitter
and 'immorality' of the city. Similarly, the occasional serious
gestures he made to elements of the modern world (such as the
testimonial he gave to the Singer sewing-machine company
and the personal closeness he maintained with some industrial
tycoons in India) prompted many of his political heirs to believe
that he took an instrumental view of everything modern. They
tried therefore to realize Gandhi's ethical goals through modern
statecraft and social engineering. Others took such gestures to
be final proof that Gandhi was a flawed, unconscious modernist. They saw Gandhi's ethics as an aspect of an unrealizable
but venerable utopia and his concept of a decent society as an
individual oddity. Jawaharlal Nehru, one of the most creative
among such Gandhians, expressed his reservations about the
Gandhian vision with great charm. 'Bapu,' he said, 'you are
infinitely greater than your little books.'[41] (Instead of viewing
this as an instance of Nehru's shallow, dated modernism, one
could perceive it also as one instance of the inability of many
of Gandhi's heirs to appreciate the extent to which Gandhi was
to remain contemporary to the post-modern generations. It is
remarkable how hostile progressives often are to the progressives bypassed by time. Nehru thought that he would survive
as a modernist, and as a Gandhian succumb to the ravages of
time. As it happens, Nehru lives even in the modern world

[40] Mao, who otherwise sinified Marx and sought to use Marxism for the needs
of the Chinese culture, nevertheless ended up by trying to rebuild Chinese culture
according to his version of Marxism and to defeat Confucius. His successors
with their theory of the four modernizations have only strengthened a vector
that was always latent in Mao.

[41] Quoted in Mahadevan, *Dvija*, p. 144.

mainly as a Gandhian; his modernism, like his version of socialism, is as pathetically orphan as last year's fashions.)

Such attempts to contain Gandhi by praising his ethics and
devaluing his thought, though common enough, flout the first
principle of Gandhi's critique of modernity. Gandhi rejected
modernity not on the grounds of ethics alone but also of knowledge; the two reasons were intertwined. He would have been
horrified by any plea for non-violence which was not informed
with the belief that non-violent methods were superior to
Machiavellism, morally *and* strategically.[42] Traditional technology, too, was for him an ethically *and* cognitively better
system of applied knowledge than modern technology. He rejected machine civilization, not because he was a saint making
occasional forays into the secular world, but because he was a
political activist and thinker with strong moral concerns.[43] That
is why attempts to contextualize Gandhi by referring to the
relatively humane conduct of the British in India fail after a
point. He was not a saint whose methods accidentally succeeded
under a benign regime which recognized his saintliness. Rather,
the methods had a built-in awareness of the nature of man-
made suffering in our times: they had been evolved in a proper
police state which had racism as its declared ideology, South
Africa. They succeeded, at least on one occasion, against the
Nazi state in war-time Berlin;[44] and they might have succeeded
in India, as Rajiv Vora and Rakesh Bharadwaj show, not due
to the civility of the British authorities but despite its absence.[45]

[42] This is to deny that Gandhi could be shelved as a man of religion, not that
ethics was not a salient factor for Gandhi. See note 43 below.

[43] Throughout this essay I have stressed the importance of Gandhi as a thinker,
underplaying to some extent his importance as an activist. This is because he
suffers from a fate similar to Leo Tolstoy's. Tolstoy is supposed to be a great
novelist and a bad thinker; Gandhi is supposed to be an outstanding Indian
politician and a bad thinker. As a result, many have tried to build a Gandhi
entirely out of the story of his life. It is only recently that his thought is being
given some direct critical attention. Cf. Isaiah Berlin, 'Tolstoy and Enlightenment', in *Russian Thinkers* (Harmondsworth: Penguin, 1978), pp. 238–60.

[44] Gene Sharp, *The Politics of Nonviolent Action* (Boston: Porter Sargent, 1973),
Vol. 1, pp. 87–90.

[45] Rajiv Vora and Rakesh Bharadwaj, 'Uncivilized Civility and Civilized
Satyagraha', *Gandhi Marg*, 1981, *3*, pp. 266–79.

One might even say, it was Gandhi's technique which elicited from his opponents a relatively humane response.

On the other hand, of the many cultural critics of the modern West, Gandhi was one of the few who never used modern scientific rationality to justify alternative cultural visions. He judged a culture in terms of what he saw as the eternal values, needs and reasons, not in terms of the culture's compatibility with modern science in the fashion of James Jeans or Fritjof Capra. Cultural myths were central to his concept of social life and political mobilization, yet he never consciously used them as instruments. Nor did he see the myths as valuable because they were rational or scientific in the sense of modern structural anthropology. The myths were seen as ethically and cognitively valid—or invalid—in themselves. Indeed, rationality and science were sometimes judged in terms of these myths. Above all, unlike many cultural conservatives and mystics, Gandhi never sought to bypass the modern experience. He explicitly took into account the experience and integrated it within the traditionalism he espoused.[46]

His utopia he called *Ramarajya*, the kingdom of the mythical Kshatriya king Rama, who was banished to the wilderness and fought, at the head of a primitive pastoral army, the demon-king Ravana. Ravana, his antonym, was not fully evil, not even in the Ramayana. Though depicted as a *brahmarakshasa*, the most dangerous form of *rakshasa* or demon, Ravana was also recognized as a learned, urbane Brahman with enormous technological and martial skills and with a clear sense of power, achievement and realpolitik. According to some readings of the

[46] During the post-Partition holocaust in India, Gandhi looked for, of all persons, M. N. Roy—ultra-rational, materialist, wedded to modern science and technology—as a possible ally in his battle against religious animosities. The M. N. Roys made sense to him and they were functional parts of a Gandhian vision. Gandhi was a traditionalist because he borrowed elements from modernity, without being defensive about them, and fitted them within the traditional Indian worldview. The modern is one who either rejects traditions or alternatively accommodates them within the modern worldview. This issue has been discussed in Ashis Nandy, 'Cultural Frames for Social Transformation: Credo', in B. C. Parekh and Thomas Pantham (eds.), *Forms of Political Discourse* (New Delhi: Sage, in press).

Puranic texts, he fought Rama not only because of his *hubris* or as a result of ordained fate, but on account of what he thought were the modern demands of pure politics and progress. On his part, Rama fought a moral battle to defeat Ravana not because Ravana was a satanic figure who represented the totality of evil, but because he, Rama, represented as a godhead a mix of good and evil which, unlike the mix of good and evil Ravana projected, included a larger vision and a better balance among the traditional ends of life. The Ramayana did not reject Ravana intuitively, mechanically or purely ethically. He was considered, given due respect and then rejected as an unacceptable design of a person.

This was how Gandhi linked up his criticism of the dominant contemporary culture of politics with his concept of post-contemporary, ethical politics. In the process, he reformulated the modern world in traditional terms to make the crisis of modern times meaningful to his traditional society. He updated Indian culture, giving it a sharper and more contemporary sense of evil, and making it a holistic alternative to modernity. To call this alternative 'Indian' is of course to ignore the design of bicultural living and the alternative universalism Gandhi offered. Gandhi's was not an understanding of the West that was alien to the West, nor was his rejection of modernity the gut-reaction of a Luddite. He was one of the few non-Westerners who had carefully read and digested the relevant Western experience and he was one of the very few among the third world's nationalist leaders to see the full implication of the West's Faustian compact with modernity.

He formulated his understanding in exactly those terms, as much as an insider as an outsider.

Index